PACK OF TWO

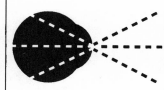
This Large Print Book carries the
Seal of Approval of N.A.V.H.

PACK OF TWO

The Intricate Bond
Between People and Dogs

CAROLINE KNAPP

Thorndike Press • **Thorndike, Maine**

Published in 1999 by arrangement with The Dial Press, an imprint of Dell Publishing, a division of Random House, Inc.

Thorndike Large Print® Americana Series.

The tree indicium is a trademark of Thorndike Press.

The text of this Large Print edition is unabridged.
Other aspects of the book may vary from the original edition.

Set in 16 pt. Plantin by Al Chase.

Printed in the United States on permanent paper.

Library of Congress Cataloging in Publication Data
Knapp, Caroline.
    Pack of two : the intricate bond between people and dogs / Caroline Knapp.
      p.   cm.
    Originally published : New York, NY : Dial Press, 1998.
    ISBN 0-7862-1726-X (lg. print : hc : alk. paper)
    1. Dogs — Behavior — United States — Anecdotes.
    2. Dogs — Training — United States — Anecdotes.
    3. Dog owners — United States — Anecdotes.
    4. Human-animal relationships — United States — Anecdotes.
    I. Title.
    [SF433.K54 1999]
    636.7′088′7—dc21
                                             98-46085

For my mother

AUTHOR'S NOTE

Names and identifying details of some owners and their dogs have been changed at the owners' request.

CONTENTS

	Prologue	9
1.	The Color of Joy	11
2.	Fantasy Dog	29
3.	Nineties Dog	65
4.	Bad Dog	93
5.	Inscrutable Dog	135
6.	Our Dramas, Our Dogs	175
7.	Family Dog	215
8.	Surrogate Dog	249
9.	Therapy Dog	291
	Epilogue	334
	Source Notes	346
	Acknowledgments	363

PROLOGUE

I am sleeping with a dog. It's two A.M., maybe three, and I wake up to find her small, slender face a foot or two away from mine, eyes fixed on me. This is the dog's silent vigil, and like most of her behaviors, it has great purpose. If the vigil fails, if the dog stares at me and I don't stir, she will employ an alternative strategy, licking my hands and face until I comply. This is dog code; it means: *Wake up and let me under the covers, please,* and like so many aspects of life with a dog, it has become a ritualized event that seems to please us both deeply. I wake up, say, "Okay," then pull the covers back. She creeps in headfirst and curls against my stomach, her nose pressed against my knees. And we both sigh. I feel so profoundly contented at those moments, snuggled under the covers with the dog, that I sometimes resist the pull back toward sleep and I lie there for a few minutes, trying to soak up the feeling: intimacy.

I have fallen in love with my dog. This happened almost accidentally, as though I woke up one morning and realized: Ooops!

I'm thirty-eight and I'm single, and I'm having my most intense and gratifying relationship with a dog. But we all learn about love in different ways, and this way happens to be mine, through a two-year-old, forty-five-pound shepherd mix named Lucille.

1

THE COLOR OF JOY

Imagine a scaled-down, delicately boned German shepherd dog, black and gray and tan instead of black and sable like a purebred, her face the color of ink with a faint gray mask. This is Lucille, a most ordinary-looking dog. She does have some exceptional features — her two forelegs are white, one halfway up from the paw, the other about a quarter of the way, which create the impression that she is wearing ladies' gloves; there is also the tiniest bit of white mixed into the fur at her chin, which makes her look vaguely like Ho Chi Minh if you catch her at the right angle. But for the most part, she is the kind of dog you might see pictured in the dictionary under "mongrel" or, if you happen to own a more politically correct edition, "mixed breed." Unremarkable, in other words, but no matter. When you study a dog you love, you find beauty in every small detail, and so it is with Lucille: I have become enchanted by the small asymmetrical whorls of white fur on either side of her chest, and by her tail, which she carries in a high confident curve, and by her eyes, which are watchful and intelligent, the color of chestnuts. I am in love with the dog's belly, where the fur is fine and soft and tan, and I am charmed by her jet black toenails, which stand out against the white of her front paws as though they've been lacquered, and I am deeply admiring of her demeanor, which is elegant and focused and restrained. I seem to spend a great deal of time just staring at the dog, struck by how mysterious and beautiful she is to me and by how much my world has changed since she came along.

Before you get a dog, you can't quite imagine what living with one might be like; afterward, you can't imagine living any other way. Life without Lucille? Unfathomable, to contemplate how quiet and still my home would be, and how much less laughter there'd be, and how much less tenderness, and how unanchored I'd feel without her presence, the simple constancy of it. I once heard a woman who'd lost her dog say that she felt as though a color were suddenly missing from her world: the dog had introduced to her field of vision some previously unavailable hue, and without the dog, that color was gone. That seemed to capture the experience of loving a dog with eminent simplicity. I'd amend it only slightly and say that if we are open to what they have to give us, dogs can introduce us to several colors, with names like wildness and nurturance and trust and joy.

I am not sentimental about dogs, my passion for Lucille notwithstanding. I don't share the view, popular among some animal aficionados, that dogs are necessarily higher beings, that they represent a canine version of shamans, capable by virtue of their wild ancestry or nobility of offering humans a particular kind of wisdom or healing. I don't think that the world would be a better place if everyone owned a dog, and I don't think that all relationships between dogs and their owners are good, healthy, or enriching. "Dogs lead us into a kinder, gentler world." Honey Loring, a woman who runs a camp for dogs and their owners in Vermont, said this to me about a year after I got Lucille, a statement that struck me as rather flip. No: dogs lead us into a world that is sometimes kind and gentle but that can be frightening, frustrating, and confusing, too. Dogs can be aggressive and stubborn and willful. They can be difficult to read and understand. They can (and do) evoke oceans of complicated feelings on the part of their owners, confusion and ambivalence about what it means to be responsible, forceful (or not), depended upon. They can push huge buttons, sometimes even more directly than humans can, because they're such unambiguous creatures, so in-your-face when it comes to expressing their own needs and drives: if you've got problems asserting authority, or insecurities about leadership, or fears about being either in or out of control, you're likely to get hit in the face with them from day one. In my view, dogs *can* be shamanistic, can be heroic and gentle and wise and enormously healing, but for the most part dogs are dogs, creatures governed by their own biological imperatives and codes of conduct, and we do both them and our relationships with them a disservice when we romanticize them. Writes Jean Schinto, author of *The Literary Dog*, "To deny dogs their nature is to do them great harm."

That said, I also believe that dogs can — and often do — lead us into a world that is qualitatively different from the world of people, a place that can transform us. Fall in love with a dog, and in many ways you enter a new orbit, a universe that features not just new colors but new rituals, new rules, a new way of experiencing attachment.

Everything shifts in this new orbit, sometimes subtly, sometimes dramatically. Walks are slower: you find yourself ambling up a city street instead of racing to a destination, the dog stopping to sniff every third leaf, every other twig, every bit of debris or detritus in your path. The clothes are different: pre-dog, I used to be very finicky and self-conscious about how I looked; now I schlepp around in the worst clothing — big heavy boots, baggy old sweaters, a hooded down parka from L.L. Bean that makes me look like an astronaut. The language is different, based on tone and nuance instead of vocabulary. Even the equipment is new and strange: you find yourself ordering unthinkable products from the Foster & Smith catalog (smoked pigs' ears, chicken-flavored toothpaste), and you find your living-room floor littered with sterilized beef bones and rawhide chips and plastic chew toys and ropes and balls, and you find your cupboards stocked with the oddest things — freeze-dried liver cubes, tick shampoo, poop bags.

The internal shifts are bigger, sometimes life-altering. When you speak to people about what it's like to live with a dog, you hear them talk about discovering a degree of solace that's extremely difficult to achieve in relationships with people, a way of experiencing solitude without the loneliness. You hear them talk about the dog's capacity to wrest their focus off the past and future and plant it firmly in the present, with the here-and-now immediacy of a romp on the living-room rug or a walk in the woods. You hear them talk about joys that are exquisitely simple and pure: what it's like to laugh at a dog who's doing something ridiculous, and how soothing it is to sit and brush a dog's coat, and how gratifying it is to make a breakthrough in training a dog, to understand that you're communicating effectively with a different species. Above all, you hear them talk about feeling *accepted* in a new way, accompanied through daily life and over the course of years by a creature who bears witness to every change, every shift in mood, everything we do and say and experience, never judging us when we falter or fail.

Of course, not everybody gets this. Fall in love with a dog, and among non–dog people, you will see eyebrows rise, expressions grow wary. You'll reach into your wallet to brandish a photograph of a new puppy, and a friend will say, "Oh, no — not *pictures.*" You'll find yourself struggling to decline an invitation for a getaway weekend — to a hotel or a spa or a family home, somewhere dogs are not permitted — and you'll hear the words, "Just kennel the dog and come on down." You'll say something that implies profound affection or commitment, and you'll be hit with the phrase, dreaded words to a dog lover, "Oh, please, it's *just a dog.*"

More commonly, you'll get vacant looks. A married friend who lives in Los Angeles, someone I don't often see, was in town recently and came to my house for dinner. At one point, sitting in my living room, he looked around and asked me, "So what's it like living alone? What's it like getting up alone every morning and coming home every night to an empty house?" I was on the sofa, Lucille curled against my thigh. I pointed to the dog and said, "But I'm not alone. I have her." He said, "Yeah, but . . ." He didn't finish the sentence, but he didn't have to. He meant: Yeah, but a dog isn't the same as a human. A dog doesn't really count.

Attitudes like this can make dog lovers feel like members of a secret society, as though we're inhabiting a strange and somehow improper universe. Not long ago, over dinner with a non–dog friend named Lisa, I started talking about Lucille, and how important her presence had been to me during the breakup of a long-term relationship. The breakup was recent, and it was long and painful and scary, as such things are, and at one point I said quite candidly,

"I'm not sure I would have been able to face the loss if I hadn't had the dog."

This seemed like a perfectly reasonable statement to me — I tend to take my attachment to her for granted these days, as a simple and central fact of life — but Lisa's eyes widened a little when I said it. She said, "Wait a minute. You're scaring me."

Scaring her? I looked at Lisa, aware of a sudden sense of dissonance, as though I'd just exposed too much. It was an uh-oh feeling: Uh-oh, she doesn't live in that world, she probably thinks I'm wacko.

So I took a deep breath and tried to explain. This is a complicated task, trying to describe how a relationship with a dog can be healthy and sustaining and rich. It's hard even trying to explain that the attachment does, in fact, qualify as a relationship, a genuine union between two beings who communicate with, respect, and give to one another. Unless you fall back on the one or two pat explanations we routinely trot out in order to explain the canine place in the human heart — dogs give us unconditional love, dogs are "good companions" — it's hard to talk about loving a dog deeply without inviting skepticism. A lot of people, quite frankly, think intense attachments to animals are weird and suspect, the domain of people who can't quite handle attachments to humans.

So there was a good deal I didn't tell Lisa. I didn't talk about what a central force in my life Lucille has become in the years since I acquired her. I didn't talk about how I basically structure my life around the dog, organizing the day around the morning walk, the noon walk, the evening outing. I didn't tell her how much I think about Lucille, how much I hate leaving her alone when I have to go out, how I've either written off or vastly reduced my involvement in activities that don't include her — shopping, movies, trips that involve air travel. I didn't use words like joy or love or affection, although it's safe to say that Lucille has given me direct and vivid access to all those feelings.

Nor did I tell Lisa how much I *need* the dog, which might have been the most honest thing to say. Lucille came into my life in the aftermath of a period of enormous upheaval. In the three years before I got her, both my parents had died, my father of a brain tumor and my mother of metastatic breast cancer. Eighteen months to the day before I got her, I'd quit drinking, ending a twenty-year relationship with alcohol, and opening up a third abyss in my life. So I was wandering around at the time in a haze of uncertainty, blinking up at the biggest questions: Who am I without parents and without alcohol? How to make my way in the world without access to either? How to form attachments, and where to find comfort, in the face of such daunting vulnerability? Lucille has been a fundamental part of my answer to those questions: in her, I have found solace, joy, a bridge to the world.

But I didn't go into all that with Lisa. Instead, I used safe descriptions, clinical terms. I talked about loneliness, and how Lucille's presence had helped ease the fear and emptiness that accompany a major breakup. I gave the dogs-as-pack-animals speech, explaining how dogs' need for social structure really does turn them into family members of sorts, highly relational creatures who look to their owners for leadership and guidance and companionship. I talked about what a comforting presence she is, how much pleasure I get out of walking in the woods with her, watching her play, even just sitting beside her while she's curled up on the sofa at home.

Lisa seemed to respond positively enough to this line of thought — "right," she said at one point, "they are good companions" — but I was aware as I talked of a gnawing frustration, a sense of my own compulsion to hold back when I talk about my dog and to offer up what's in effect a watered-down and fairly stereotypical view of the attachment: dog as man's best friend, dog as loyal and faithful servant. There are elements of truth to that view — dogs can be wonderful friends, they can be enormously loyal and faithful creatures — but those factors represent only one part of the picture, a limited and really rather arrogant fragment that concerns only the way dogs serve us, not the ways we serve them or the ways we serve each other. Finally, I shook my head and said to Lisa, "You know, it's been really important to me to learn not to pathologize my relationship with Lucille. People have very powerful relationships with their dogs, and that doesn't mean they're crazy, or that they're substituting dogs for humans, or that they're somehow incapable of forming intimate attachments with people. It's a different *kind* of relationship, but it's no less authentic."

Alas. Lisa looked across the table and said, "You're still scaring me."

Dog love, popular wisdom suggests, should be limited love. Let on the depth of your true feelings about a dog — how attached you are, how vital the relationship feels — and you risk being accused of any number of neuroses: you're displacing human love onto the animal, which is perverse; you're anthropomorphizing, which is naive and unsophisticated; you're sublimating your unconscious wish for a baby or a spouse or a family into the dog, which is sad and pathetic. Children are allowed to harbor deep affection for dogs: that's seen not only as cute and normal but as morally acceptable, as caring for a pet can teach a child about compassion and responsibility, even about loss, given a dog's relatively short life span. The elderly and the infirm are permitted some degree of attachment, too, thanks in recent years to widescale acceptance of the use of therapy dogs in settings like nursing homes and hospitals. But the rest of us are expected to keep our feelings about dogs somehow contained and compartmentalized, in the box labeled "Just a Dog." And if we don't — well, as my friend Lisa said, we're a little scary.

In fact, more than one third of all Americans live with dogs today — by most reliable estimates, that's about 55 million dogs — and it's safe to say that a good number of us don't contain or compartmentalize our feelings nearly so effectively. Suspect though dog love may be in the public eye, Americans are in the midst of a veritable love affair with dogs: we're spending more money on our dogs than ever before (the average owner can expect to shell out a minimum of $11,500 in the course of a dog's life); we're indulging them with an ever more elaborate range of goods and services (doggie day care, doggie summer camp, gold-plated Neiman Marcus doghouses); and in many respects we're treating them far more like members of the human pack than like common household pets. Depending on which study you look at, anywhere from 87 to 99 percent of dog owners report that they see their dogs as family members, figures that are certainly borne out by behavior. The American Animal Hospital Association conducts an annual survey of pet owner attitudes. In 1995, 79 percent of respondents reported that they give their pets holiday or birthday presents. Thirty-three percent said they talk to their dogs on the phone or through an answering machine when they're away. If they were stranded on a desert island and could pick only one companion, 57 percent of owners said they'd choose to be marooned with the dog rather than a human. A more telling number: the following year, 48 percent of female respondents reported that they relied more heavily on their pets than on their partners or family members for affection.

I understand the temptation to pathologize such behavior, or at least to poke fun at it (dogs in birthday hats?), but I don't believe that dog owners are unilaterally engaged in displacement, sublimation, or rampant anthropomorphism. Nor do I see this apparent depth of attachment as a sad commentary on contemporary human affairs. This is another common view, that people turn to pets for love and affection by default, because "real" (read: human) love and affection are so hard to come by in today's fractured, isolated, alienating world. I think there is a kernel of truth to that — we live in lonely times, and dogs can go a long way toward alleviating loneliness — but I think the more important truth has to do not with modern culture but with dogs themselves, and with the remarkable, mysterious, often highly complicated dances that go on between individual dogs and their owners.

That dance is about love. It's about attachment that's mutual and unambiguous and exceptionally private, and it's about a kind of connection that's virtually unknowable in human relationships because it's es-

26

sentially wordless. It's not always a smooth and seamless dance, and it's not always easy or graceful — love can be a conflicted, uncertain experience no matter what species it involves — but it is no less valid because one of the partners happens to move on four legs.

"Love is love. I don't care if it comes from humans or from animals: it's the same feeling." Paula, a forty-seven-year-old children's book author who lives in Los Angeles with three Maltese dogs, said this to me with such simple candor the words stuck with me for days. She continued: "When I'm feeling bad or thinking about something I can't handle, I pick up my dogs and it helps for that moment. It may not be the perfect relationship we all hope to have with a human, but it's a relationship. And love is love."

Indeed. Just this morning, I came into the house after being out for an hour or so and found Lucille nestled in a corner of the sofa, her favorite spot when I'm away. She didn't race across the room to greet me — she's sufficiently accustomed to my comings and goings by now that she no longer feels compelled to fly to the door and hurl herself onto me as though I've just returned from the battlefield — but when I came into the room and approached her, her whole body seemed to tighten into a smile: the pointed ears drew flat back, the tail thumped against the sofa cushion, the eyes gleamed, the expression took on a depth and clarity that suggested, *Happy; I am completely happy.* A friend says her dog seems to wake up every morning with a thought balloon over her head that says, *Yahoo!* That was precisely the look: *All is right with the world,* it said, *you are home.* I crouched down by the sofa to scratch her chest and coo at her, and she hooked her front paw over my forearm. She gazed at me; I gazed back.

I have had Lucille for close to three years, but moments like that, my heart fills in a way that still strikes me with its novelty and power. The colors come into sharp focus: attached, connected, joyful, *us.* I adore this dog, without apology. She has changed my life.

2

FANTASY DOG

The puppy — Lucille at ten weeks — is stalking an ant. She creeps along the edge of the fence on my patio, head tucked low, neck stretched forward, her step delicate and calculated and silent. I can't see the ant so I'm not exactly sure what she's doing, but I can see the intensity of her focus. She stops, her gaze fixed at something about two feet ahead of her, and then her body tenses, and then she *pounces:* lunges forward, lands, and stops, front paws planted on the brick, hindquarters raised, little curl of a tail swishing behind her.

I watch this, and I smile. Smile and smile and smile. In the course of two weeks, this creature has crawled into a corner of my heart and gotten lodged there, permanent occupancy. I look at her and sometimes I have to clench my teeth to keep from grabbing her and squishing her, she so delights me. Where did you *come* from? I gaze at her, wondering this. How did I end up with *you?*

Good questions; I'm still asking them today. I stumbled into the world of dogs with major blinders on, just kind of woke up one day with this animal in my house. This is not an exaggeration, either — I've acquired toasters with more deliberation than I acquired Lucille. One fantasy, one animal shelter, fifty bucks, puppy. In retrospect, I'm also astonished by this: I'm not by nature a spontaneous person, and I'm certainly not a rash one, so the idea that I'd just go out one afternoon and come home with a live animal seems completely out of character. But fate has an uncanny way of giving you what you need, presenting you with the right lessons at the precise moment you're ready to learn them. At the time I needed to learn a lot — about connection and closeness and safety — and something deep inside whispered, *A dog, you need a dog,* and I was lucky enough, or open enough, to listen.

A few weeks before I got Lucille, I'd been sitting at a table outside a café in Cambridge with my friend Susan, watching a trio of people and a dog at a nearby table. The dog was sweet-looking in a mangy way, a medium-size mixed breed with intelligent eyes, and he sat next to his owner with that patient, contented look that some dogs seem to wear all the time: ears relaxed, eyes bright, mouth partway open in a mild pant that looked like a smile. At one point the dog stood up and started to wander away from the table. The owner whistled lightly, and the dog stopped, looked over his shoulder, then trotted back to the table and rested his head on the owner's knee. The owner gave him a soft pat, then returned to his coffee. The two looked utterly at home together, man and dog.

I watched this with vague envy, the way I might watch a couple holding hands. "I want a dog," I said to Susan, nodding in the pair's direction. "I'm thinking about getting a dog."

This was the first time I'd said the words aloud, although in the days and weeks before I got Lucille, I'd been aware that dog thoughts had been circling in the back of my mind: dog fantasies, a kind of low-grade puppy lust.

A dog. I grew up with dogs, first a lovely, loyal, beautifully behaved Norwegian elkhound named Tom, who died when I was in high school, and then a lovely, utterly disloyal, terribly behaved elkhound named Toby, who died when I was thirty, just a few years before my parents' deaths. Elkhounds are a burly breed with thick black and gray coats, curly tails, and wolfish faces; they're charming dogs, intelligent and lively, but they belonged primarily to my mother, and I rather took them for granted as a kid — nice

33

dogs, friendly, and largely peripheral to me. Still, when you're raised in the presence of dogs, you tend to align yourself with them in some fundamental internal way. Studies by James Serpell, professor of animal welfare at the University of Pennsylvania, suggest that people exhibit a fair amount of species loyalty when it comes to acquiring pets: if you grew up with a dog, you're likely to end up with a dog yourself as an adult, while if you grew up with cats, you're more likely to stick with felines. This, at least in part, is what turns us into "dog people" or "cat people," and the phenomenon certainly applied to me: I grew up simply assuming that at some point I'd end up with a dog of my own.

Perhaps because I was never charged with taking care of the dogs as a kid, I also had lots of ideals about them, a passel of bright images about what living with a dog might be like. Never mind the way Toby used to torture my mother by escaping from our yard at every available opportunity and hightailing it to the local Dunkin' Donuts, where he'd loiter and beg for tidbits; never mind the times he'd disappear at our summer house on Martha's Vineyard, often returning days later with little notes tied to his collar. ("Your dog showed up at our cocktail party; we gave him a drink and sent him on his way.") I remembered, instead, the way Toby used to make my parents laugh. (They always got a chuckle out of those notes, the annoyance factor aside.) I remembered the way Tom would lie by the fire during the evening cocktail hour and stare at my dad, whom he adored. I remembered the way my mother, a markedly undemonstrative woman most of the time, used to reach down and pet the dog, scratch his chest until he zoned out with contentment, eyes at half mast. My associations had to do with loyalty and companionship, with affection and nurturance, with simple tactile joys.

My fantasies, meanwhile, had to do with attachment, and with longings that felt both inchoate and vast. At the time of that coffee with Susan, my future was looming before me like a dictionary definition of uncertainty. I was still somewhat dazed from my three-year bout with loss: orphaned and raw and fearful, struggling to figure out how to have a life without being drunk all the time. In the midst of that task, the external elements of my life were in flux, increasingly unsettled and ill formed. Earlier that week, I'd put the finishing touches on the manuscript for my last book, a memoir of my experience with alcholism, and then, days later, I'd quit my full-time job, a position at a newspaper in Boston I'd held for seven years. So I was poised at the edge of the abyss labeled "self-employment," absolutely unsure how I'd fill my days. My personal life felt just as open-ended and even more daunting. I was deeply ambivalent about my then-boyfriend, a man of uncommon kindness and generosity named Michael. We'd been together for five years, and he'd seen me through the worst times — the deaths, the decision to quit drinking — and I still couldn't for the life of me figure out what to do with him, whether to marry him or up and run. A month before I got Lucille, we'd gone looking at houses for sale, fantasizing about creating a home together. I'd stand there in a big old Victorian in Cambridge, and I'd imagine us cooking spaghetti in the kitchen, or tending to the yard. And I'd also think: Is this place big enough so that I can be way over here and he can be way over *there?* For our entire relationship, I'd been keeping Michael in a box, and I held the only key: I was the gatekeeper of intimacy, in charge of deciding how close he got, how many nights a week we spent together, how often we had sex. This was beginning to make both of us nuts: the manipulation of distance and the imbalance of power and the sense that time was moving forward while our relationship stood still, and I'd slouch into my therapist's office week after week and howl about my ambivalence. I can't do this anymore; I can't leave him, I can't stay; tell me what to *do*.

I couldn't talk to my twin sister about this, which distressed me greatly, she being my closest family member and the other key figure in my life. In the midst of all my indecision, she'd left her husband, fallen in love with another man, and was talking about moving to a new part of the state, and every time I talked to her, I had the feeling that she'd taken off and landed on some new planet, without leaving behind directions. Our conversations felt stilted and disconnected: we'd talk on the phone, and she'd seem utterly absorbed in her new relationship, and every time we hung up, I'd be aware of a sensation that's dogged me since childhood, that she'd surged forward in her life and left me behind.

So life had this unmoored quality, full of voids and barely acknowledged yearnings, and if I'd made a list of things I wanted desperately at the time, it would have included the most elusive items. Love without ambivalence. Family members who won't leave.

Intimacy that's not scary, that doesn't require a lot of anesthesia.

I suppose that's why the sight of that man and his dog appealed to me so: it spoke to that list, ticked off promises of possibility. A warm, uncomplicated connection. A singular attachment. Ease on a leash. I sat and watched. I wanted what they had. And so I said it aloud: "I'm thinking about getting a dog."

Susan is a very intuitive person who understands that I have difficulty with the concept of indulgence. She looked at the dog and his owner, then looked back at me. Her eyes lit up a bit, as though I'd hit upon something inexplicably perfect. "A dog," she said, turning the word over. "I like it. Caroline and her dog."

As it turns out, my fantasies were utterly typical, as classic as they come. Ask ten people why they want a dog, or why they got one, and you will get ten variations on the same theme: dog equals love. More to the point, dog equals a very specific brand of love: a warm-and-fuzzy variety, pure and simple, low-maintenance and relatively risk free. A dog will return us to more idyllic times, to summer afternoons we spent romping with the family dog as children. A dog will do for us what Lassie did for Timmy, provide constancy and protection and solace, our very own saint in the backyard. A dog will curl at our feet and gaze up at us adoringly, will fetch our paper in the morning and our slippers come nightfall, will serve us and love us without question or demand. This is Walt Disney love, rose-colored and light and tender, and the wish for it lurks within every human soul, the dog owner's being no exception.

People tend to be surprisingly vague when they talk about why they decided to get a dog — I've been struck by this countless times. You'll ask, "Why a dog?" and "Why *then?* Why did you get a dog at that particular time in your life?" and you'll often get a highly generic response, or a pragmatic one. A mother of two boys echoes the parental refrain when she tells me, "We got the dog for the kids." A retired schoolteacher in Washington cites the fitness-and-activity rationale: "I wanted a pet that would get me outdoors, give me a reason to get out of the house for walks." In interviews dozens of dog owners in varying categories — single women, single men, married couples — trotted out variations of an even less specific theme: I don't know, I just like dogs. I grew up with dogs and I always wanted one.

Those are all valid, perfectly reasonable responses — dogs can be great playmates for children; dogs will get you out of the house for walks; dogs are, or at least can be, eminently likable — but I think people tend to be vague because getting a dog can be such an intensely personal matter, all tied up with that Disney ideal, and the very personal fantasies and yearnings that lurk behind it. Even the most pragmatic rationales contain deeper hopes. The family dog offers the promise of stability: the house in the suburbs, the picket fence, the golden retriever in the back of the station wagon. The outdoor dog offers the promise of fidelity and companionship: the trusted Lab trotting by your side. The statement "I grew up with dogs" suggests a longing for the purity and innocence of childhood bonds, a wish for simpler times and less cluttered relationships. Dogs strike deep chords in us, ones that are bolstered by the individual experiences of childhood, by the culture at large, and by history. Humans have lived with dogs for fourteen thousand years, after all, and our ideals about them are deeply rooted, bred into the dog-loving portion of the population as surely as the instinct to chase prey has been bred into hounds. We've rhapsodized about dogs in literature and poetry, filled centuries worth of canvases with images of his nobility and strength, celebrated his loyalty and steadfastness in film; from Odysseus' Argus, who waited longer than Penelope, to the long-distance loyalty of Lassie, we have exalted the dog for attributes all too often absent from human affairs.

We have done so, moreover, with remarkable consistency, our ideals about what dogs are like and what they can provide remaining more or less the same for centuries. Times when I wonder if my feelings for Lucille are over the top, when I question whether my affection for her is somehow excessive, I remind myself of the volumes that have been written about dogs, of the persistence and constancy of their place in our lives. In an 1862 letter to a friend, poet Emily Dickinson summed up in two quick sentences the precise qualities of trust and camaraderie the modern dog owner so values and enjoys: "You ask of my companions. Hills, sir, and the sundown, and a dog large as myself that my father gave me. They are better than human beings because they know but do not tell." In 1930 Thomas Mann published *A Man and His Dog*, a 258-page love letter to his hound dog, a "wondrous soul" named Bashan, which captures, among other things, the joy a human feels at the sight of a dog bounding through the woods, "sagacious, vigiliant, impressive, with all his faculties in a radiant intensification." In 1940 French psychiatrist Marie Bonaparte published *Topsy*, a tribute to her chow chow, whom she describes in terms as loving and familial as any you will hear today: "Topsy is my friend," she wrote, "my friend, who, in this, different from my grown-up children, does not ask to leave me. . . . Topsy lives and breathes in a radius ten yards around me, and cries to rejoin me as soon as I move a few steps away from her. Dogs are children that do not grow up, that do not depart."

Dogs as cherished companions, dogs as aesthetic wonders, dogs as devoted family members: those themes predate my own preoccupation with Lucille by centuries, and it's probably no surprise that when I set out to the pound to find her, I carted the hopes and longings they represent right along with me.

Of course, I didn't set out to fill those longings deliberately. Certainly I didn't sit back one day and think: Gee, I've lost both parents and quit drinking and my life is full of gaping holes; guess it's time to get a puppy. Instead, I woke up on a Sunday morning in August, an unplanned day looming before me, and I thought: What the hell. Maybe I'll go to the pound and just look. I wasn't sufficiently wedded to the idea of a dog to launch a full-scale investigation of breeds and breeders, and I had an instinctive aversion to pet store puppies (a good instinct, as it turns out, since an alarming proportion of pet store dogs come from puppy mills, which churn them out for volume and profit and with little regard for health or breeding). I also liked the idea of rescuing a dog, so I mulled the notion over, and I got up off my sofa and headed out the door.

One thing I've noticed since I quit drinking is that a person usually has two or three sets of impulses scratching away at some internal door at any given time. If you're sober — if you're alert, and paying attention to those impulses, and not yielding to the instinct to anesthetize them — you can receive a lot of guidance about where to go, what to do next in life. Some people in AA define this as their higher power, as though there's a part of each of us, a kind of higher self, that wants to be healthy and well, that can set us on the right path if only we heed its messages. I felt that scratching that morning, as though an invisible thread were attached to my soul and tugging at me ever so gently. *Do it. Just go look.* Of course, the part of me that wants to resist a healthy impulse can be incredibly strong, so looking back, I'm still surprised that I actually got in my car and came home with a puppy. But the thread kept tugging, and I managed not to dismiss it.

I drove out to Sudbury, Mass., first, where my sister lived, and she took me to an animal shelter near her house. We didn't find anything there — the kennel consisted of about two dozen pens, a dozen on each side of a long hall, and none of the dogs struck me in quite the right way. There were some large, loud, lunging dogs who looked like too much to take on; a few small, hyper-looking dogs who seemed too high strung and yippy; and a few others who were too old, or too funny looking, or too sickly, or somehow just not what I had in mind. It's a weird feeling, walking down a hallway of rejected animals and rejecting them all over again, and the experience made me feel guilty and uncomfortable, as though I lacked some requisite degree of altruism or humanity. That feeling gnawed at me and almost turned me off the idea entirely, but when I got back to my sister's house, the thread tugged again. *Go somewhere else.* And so it was that an hour later, I walked into the Animal Rescue League in downtown Boston, equipped with the ID and documentation required to adopt a pet, ready, I guess, to find my dog.

Lucille is a remarkably serene dog, poised even as a puppy, and that struck me about her from the start. When I first saw her, she was in a small cage in a corner of the shelter, with barking, yelping, yowling dogs on all sides. A spaniel in the cage beside hers kept charging at its cage door, up and back, up and back. A large husky mix in a bigger cage barked and clawed at the door. Eyes beseeched, nails scratched against metal, I found myself not wanting to look. But there she was. Amid all the noise and chaos, Lucille was lying calmly in her cage with a pink chew toy between her front paws. She looked utterly focused on that chew toy, as though she was quite capable of entertaining herself, thank you, and didn't need to claw and clamor for attention like the others, and the sight of that appealed to me in a visceral way: it spoke to a kind of grace under pressure, a quality of endurance I suppose I was looking to cultivate in my own life.

Still, she was not my ideal of the perfect

dog, at least not in an obvious way. My aesthetic ideal tends toward the sleek and muscular: Rhodesian ridgebacks and Doberman pinschers. I like big dogs, athletic dogs, and in the days and weeks before I got Lucille, I'd formed a picture of an elegant animal with wide searching eyes and a fine coat, short-haired and suedelike and earth-toned, a dog (here's an embarrassingly telling concern) that would match my furniture. Lucille does not look like this, nor did she promise to as a puppy. An information-bearing card over her cage identified her simply as a "shepherd mix," which I've come to understand is a euphemism for We Really Have No Idea, and looking at her that first day, I couldn't tell if she'd grow up to be beautiful or homely: she had dainty paws and nice proportions for a pup, but she was a bit on the runty side, her coat a somewhat mousy brown, and she looked like she might be kind of stumpy when she grew up, her legs too short for her body. Mostly she was tiny and funny looking, like an oversized rodent. I stood there looking at her. My dog? I am not at all sure.

I crouched down at the cage and watched her. She looked up happily enough, and I poked my finger between the bars. She gave it a curious sniff. Lucille has shepherd ears,

46

and they were flopped over at the time, two small triangles pointing in either direction. Very cute, but I still didn't fall in love with her right then and there. I sat and thought: Can I *do* this?

This was one of those crossroads moments, when you understand that something potentially life-altering is in front of you, and you know you could go either way, take the plunge and upset everything or cast a vote for stasis and go home to your sofa. A woman I know named Helen first saw her dog, now a four-year-old cairn terrier whom she adores, in the window of a pet store, passing by on her way to meet a friend for lunch. She stood at the window and looked at him. She went in, looked more closely, poked her finger through the bars of the cage the same way I did. The puppy licked her hand, wagged his tail. She left, went back, left again. Then she spent the entire lunch obsessed with the idea, dragged her friend in and out of the pet store, and walked around the block four times. She still doesn't know what tipped the scales for her, but she remembers the questions: Should I? Shouldn't I? Can I? Why? Why not?

The scales in my case went up and down (and up and down again) on the matter of puppyhood, the responsibility it implied. After a lot of circling, my dog thoughts had landed on the side of caution, and I pretty much figured I'd get a mature dog, maybe a year or two old, who'd arrive house-trained and equipped with a mastery of the basic commands. Naive as that logic seems to me today, it sounded like an easy enough adjustment at the time: just let the dog in, learn its ways, incorporate it into my routines. But something about seeing a puppy, so small and undeveloped, also spoke to those yearnings of mine, to the part of me that seemed to want a deeper sort of experience, a fuller involvement, a relationship that might feel more special somehow.

Special. That's what I really wondered about: Can I do special? Can I love this creature the way she's meant to be loved? Can I get her to love me? Am I capable of forming the right sort of relationship here, creating the kind of bond I saw between the man and dog at the café that day? I tend to be such a fearful person when it comes to intimacy, so self-protective and locked into my routines, so averse to commitment. How would a small vulnerable pup affect all that?

A few minutes passed. A shelter employee came up to me while I was kneeling at Lucille's cage and told me I could take her out and play with her a bit if I filled out an adoption form first. I went into an adjoining reception area and did that; she came out a few minutes later carrying Lucille in her arms. Lucille's head rested on the woman's shoulders, and she was looking around the room in a curious, peaceful way, a picture of trust and composure. The employee handed her to me, and we settled down in a corner, me sitting cross-legged on the floor and Lucille sliding about in front of me on the linoleum. She looked calm and active and alert, happy to be out of the cage and able to sniff about freely. She sniffed here and she sniffed there, puppy nails scratching on the floor. I watched her. I debated. Can I do this? I'm not sure I can do this. Can I?

Just then, Lucille did a little canine jig, lifting her two front paws one after the other and sort of hurling her whole body forward on the floor into a play bow, one of those tiny motions that's so cute, it looks like a parody of puppy behavior. Then she squatted down and peed on the floor, a yellow puddle widening beneath her. My dog trainer once asked me why I chose Lucille — why her, out of all the other potential puppies in the universe — and that image sprang to mind: a little puppy making a big old mess. The sight struck some note of familiarity in me. No one at the shelter knew where Lucille had come from or why she'd been abandoned. She'd been left there the day before, no litter mates, no explanation, no story, and looking back, I can see that she looked exactly like I felt: unmoored; in need of care; a young female pup unattached to home or family. I suppose that's what clinched the decision for me. Her vulnerability spoke to my deepest fantasy: together we would form an attachment, create some semblance of home and family, a pack of two.

I stood up, walked over to the shelter employee, and said, "I'll take her."

There is nothing quite like a puppy to wrench your mind away from the darkness, especially if you've stumbled toward one as blindly as I did. Ten minutes in, all those voids and anxieties and existential worries of mine were shelved, knocked into distant corners by a much more urgent awareness: I was utterly, woefully ignorant about dogs, I knew *nothing at all* about raising a puppy, I could expose myself in this situation — to her, to myself, to the world at large — as even more inadequate than I already thought I was. I am a person who loathes imperfection, and I did everything imperfectly, dithered my way through the most elemental questions. Where will she sleep? The first night I was so unprepared and stupid about puppy care, I shut her in the downstairs bathroom, in a cardboard box turned on its side and padded with a blanket. Highly imperfect. From the get-go, I had no equipment for puppies, no medical understanding of puppies, no sense of what to expect from puppies, and for someone who's by nature very cautious and controlled — disciplined, orderly, deliberate — this was very imperfect indeed.

I was the blind leading the canine, and the first weeks with her were a blur of chaos and distraction: racing off to the bookstore to buy books about training, racing off to the pet store to buy a crate, racing back to the pet store because the crate was too big, finding a vet, calling every person I knew who had a dog, soaking up scraps of advice: Is it really okay to give a dog a smoked pig's ear? What's a training collar? The Monks of *who?*

Yet in the midst of all that frenzy, I was also aware of something lighter, almost wondrous: a sense of thaw, a budding and glorious sense of connection. I was struck by this from the first day, by the speed and intensity with which Lucille rang all those internal canine bells — puppy! love! fuzzy! warm! She was such a trusting thing, just eight weeks old when I took her home, weighing in at a mere twelve pounds. The shelter employees handed her over to me on a little red leash, and we walked out of the building and into the sun toward my car, and she just trotted along after me, blinking in the light and sniffing at the sidewalk. This amazed me, that this animal would so willingly follow me into the unknown.

When we got into the car, she climbed up into my lap and sat there the whole way home, pressed up against my stomach. I petted her and tried to keep my eyes on the road, and I thought about my father. When my parents got their first dog, Tom, they picked him up from a breeder in New Hampshire and took him home in the car, an hour and a half drive. Tom was just a ball of gray fluff at the time, the same age as Lucille when I got her, and terrified. He spent the whole trip curled up in my father's coat pocket, and although he always loved my mother, he *revered* my father after that, in an awestruck, slightly intimidated, approval-coveting way that I understood completely because I shared it. Driving home with Lucille, I thought about Tom, and I secretly hoped that our jaunt across town together would instill depth of feeling in that same way.

Maybe it did. Lucille followed me everywhere that first night, from kitchen to living room to bathroom, and every time I left a room, she got an anxious, alert look on her face, as though she were saying, *Where's the girl? Where'd that girl go?* Within days friends were dropping by, and cooing at Lucille, and saying, "She's so bonded to you!" "She's so attached!" and my heart would swell with a private little burst of pride: until she provided it, I hadn't fully understood how badly I'd wanted that sense of connection.

One of the remarkable things about dogs, one of the reasons humans have loved dogs so well and so long, is that they are singularly well equipped to make us feel loved in return. Descendants of wolves, highly social animals who live in well-organized groups and maintain strict adherence to the group's hierarchy, dogs appear to approach relationships in much the same way we do. They have similar needs for attachment. They bond to us. They look to us for leadership, and they are remarkably adept at figuring out what their place in our lives is to be, what we want from them. Dogs even seem to communicate in many of the same ways we do, exhibiting gestures and behaviors we can read and understand, the wagging tail connoting happiness, the bared teeth connoting anger, the canine gaze suggesting everything from adoration to alarm to guilt. Before I got Lucille, I had a rough working knowledge of the dog's wolfish heritage and a basic understanding of pack theory (translation: I knew the leader was called the alpha dog), but I had no idea how it would feel to be part of a dog's pack, to be followed like that, and bonded to, and needed. Her attachment to me, so quick and unquestioning, melted me.

Melted and astonished me. Dogs possess a quality that's rare among humans — the ability to make you feel valued just by being you — and it was something of a miracle to me to be on the receiving end of all that acceptance. The dog didn't care what I looked like, or what I did for a living, or what a train wreck of a life I'd led before I got her, or what we did from day to day. She just wanted to be with me, and that awareness gave me a singular sensation of delight. I kept her in a crate at night until she was housebroken, and in the mornings I'd let her up onto the bed with me. She'd writhe with joy at that. She'd wag her tail and squirm all over me, lick my neck and face and eyes and ears, get her paws all tangled in my braid, and I'd just lie there, and I'd feel those oceans of loss from my past ebbing back, ebbing away, and I'd hear myself laugh out loud.

Dogs are fantasies that don't disappoint. I know this sounds wildly improbable, but sometimes I look back and think the dog trotted in, sniffed out that itemized emotional shopping list I'd been carrying around, and said, *Okay, you want love without ambivalence? Intimacy? A sense of family? I can do that.* We'd sit outside on my patio for hours at a stretch, and I'd watch her wander about, sniff a blade of grass here or a weed there, and I'd feel like some chamber in my heart had been blasted open, a compartment labeled "love" that had been sealed shut for years. I'd pick her up and carry her inside, and she'd rest her front paws against my shoulder, all willing and trusting and tiny, and a wave of devotion would come over me, a feeling I don't think I'd experienced before even once. The emotional purity of those early days and weeks was like discovering a map where the varied elements and sensations of love were marked and identified, as if by colored flags.

I'd hold her on my lap and stroke her fur, and I'd feel a profound contentment and I'd think: Oh, this is the comforting part of love, as opposed to the scary part. I'd take her to the park and watch her play with other dogs, see the abandon with which she threw herself into a chase or leaped up on another dog, and I'd think: Ah, this is the joyful part, the part where I feel delight. I'd roll around on the floor with her, and scratch her belly and squeal at her in a high voice, and I'd think: Right, the part where you get to act silly. All these new-mother sensations seemed to stir and then blossom — nurturance and protectiveness and a cooing, cuddly affection — and each one astonished me with its clarity because, honestly, I wasn't sure I had them in me. "How's Lucille?" Friends would ask me this, and I'd beam and say, "Oh, great: we're in love." I'd say that in a joking way, but I also meant it, and I treasured the meaning behind it — the love I felt — like a jewel.

When you drink, you anesthetize yourself to the frightening parts of intimacy, but also to the gratifying, giddy parts, to the fun and excitement. Lucille answered a fantasy I hadn't even acknowledged I harbored, one I didn't understand was possible: to love another being in an utterly unfamiliar, sober way, with access to the full range of my own emotions.

And yet here's something else about canine fantasies: dogs can destroy them — can pee on them, chew them up, literally mangle and maim them — as quickly and readily as they live up to them. Dogs do this, of course, in very obvious ways: they bark and they howl and they defecate on your rug, and until they understand the rules, they can be like furry whirling dervishes, clawing and scraping and chewing a wide swath of destruction across your living room. I know of dogs who've eaten carpeting and drapes, hair scrunchies and Barbie heads, bicycle seats, trampolines, Christmas tree ornaments, and TV remote control units; I've heard of a ten-month-old beagle in Virginia who ate an entire box of jewelry, including the owner's diamond engagement ring, her gold wedding band, a ruby brooch, and (of course) the box; a twenty-month-old Bouvier des Flandres who chewed through a waterbed in Florida, then ate a dustpan, a toilet bowl brush, and three rolls of toilet paper in the course of one afternoon; and a pair of Jack Russell terriers outside Boston who chewed a hole through the back of their owner's couch, then took a left and dug a tunnel straight through to the sofa's side. This can be very disconcerting indeed, particularly among those of us who are new to the world of dogs and who acquire them without a working knowledge of such terms as "puppy proofing" and "crate training." You enter into the relationship with that soft-focus ideal of the devoted family pet, and instead of fetching your slippers, the dog eats them. You fantasize about the dog curled contentedly at the hearth, and then you watch in horror as he tears three feet of aluminum siding off the front of your house.

I was aware of this with Lucille from the beginning, too, aware that in the midst of all that giddy, heady puppy love, a lot of my ideals about dogs — the warm fuzzy associations — were turning out to be just that: fantasies, dreams that bore little relationship to the reality of opening your life to another species. The most basic things threw me. Two minutes after I unlocked the front door to let us in, Lucille trotted through the living room and into the kitchen, squatted down, and defecated on the tile. There was nothing inherently surprising about this — the dog was only two months old, I knew she wasn't housebroken — but the sight seemed to take whatever vague image of the warm, snuggly puppy I'd formed on our drive home and blow it right up, as if to say, See? She really isn't programmed with perfect puppy software.

Nor was she programmed with well-behaved adult dog software, which managed to jar me, too. If I'd had snapshots of my expectations in my wallet, they probably would have pictured candidates for awards in obedience: dogs coming when called, dogs trotting along in a perfect heel, dogs wearing that beloved expression of willing expectancy, awaiting further instructions. Mature dogs. *Trained* dogs.

Lucille, of course, frustrated those expectations one by one, just felled them like trees. She peed everywhere. I'd see her start to squat — in the middle of the kitchen, the living room, the hall, wherever — and I'd shriek: "No! Not there! Outside! *Outside!*" And then I'd scoop her up and swoop her out the back door, and she'd look at me like, Huh? What's the problem? She did not speak English. I'd crouch across from her on the patio and open my arms and say, "C'mere, you; come over here," and sometimes she'd head right for me and sometimes she'd just roam around, sniffing a patch of dirt here, a fallen leaf there, and I'd feel lost: language failed me with the dog, perhaps for the first time. On walks she'd be all over the place. I had no idea how much exercise a puppy required, but I assumed it was a lot, so I took her out constantly, little jaunts around the neighborhood every two or three hours. She was like a twelve-pound mosquito, flitting this way and that. She strained against the leash, and she lagged behind, and she tried to poke her nose in all manner of inappropriate places — into recycling bins, through chain-link fences, down into gutters — and whatever vague image of dog walking I had (dog following my lead; dog trotting calmly by my side) went right out the window.

And right along with it went my fantasy about dog love as easy, as simple and emotionally uncomplicated. Ignorant as I was, I think I imagined a dog would be like a large, fun version of a household cat: more interesting and relational than a feline, but not necessarily any more taxing on the psyche. Wrong, wrong, wrong. I am a person who worries about relationships, and I worried about ours from the get-go. What is she doing? Why is she sleeping so much? Is she eating enough? Is she happy, sad, scared, bored, lonely, amused? How is she adjusting? I worried about my ability to anticipate her needs (every time Lucille peed in the house, I considered the accident a personal failure), and I worried about my basic character, my capacity for giving and nurturing, and I worried about failing her in some fundamental way, scarring her for life. I worried, in short, about everything from her mental health to the consistency of her stool, and once, about six weeks in and convinced that I needed to find her the very best trainer — a top-of-the-line trainer, from the Harvard equivalent of puppy schools — I actually called up the renowned Monks of New Skete and asked them if I could drive six hours to upstate New York for a little private instruction. (They said no.)

In short order, I also entered the dark and murky world of projection, a place every dog owner visits from time to time. Lucille has the serious face and demeanor of a German shepherd dog, and I could read volumes into those sober eyes. I'd think: She's just lying there, staring at me; can she tell what an idiot I am? From day one, I harbored terrible anxiety about leaving her alone. I'd make a move toward the door — even just to whisk outside and get the paper from the front stoop — and I'd see her look up at me, and I'd read fifteen tons of alarm into that expression: Oh, God, she thinks

I'm abandoning her. Bubbling up alongside all that early joy and delight were all my deepest fears: She won't love me as much as I love her, I'm inadequate to the task of caring for her; I'll fuck this up.

I have not had an easy time of relationships. I grew up with the feeling, deep in my bones, that the world was a frightening place and that of all the scary things in it (wars, natural disasters, car crashes), the scariest thing was other people. People, with their odd, unfathomable agendas and their complicated, often insatiable needs and their mysterious, changeable ways. My family held to an aesceticism, an emotional spareness and rigidity that kept the exterior looking clean and polished, any sign of turbulence and storm carefully cloaked and hidden from view. You're angry, clench your teeth, bite your tongue. You're sorrowful or sad, you disappear behind a closed door and weep in private. You have overwhelming needs or urges or desires, you save them up and keep them to yourself, and once a week for fifty minutes at a shot, you discuss them with your therapist.

This is why I drank: I didn't know how else to deal with feelings, with the pushes and pulls and strains of a full, unfettered emotional life. Feelings? No thanks; way too scary. My mother hugged me the day I graduated from college, for the first time ever that I could recall. That same summer, a few months later, my father confessed that he'd been having an affair for ten years. So this is what I learned about emotion: very dicey stuff, difficult to express, dangerous to act on. Keep emotion hidden: from others, from yourself.

I drank and drank. The drinking accomplished two paradoxical goals, anesthetized feeling and gave me access to it at the same time. When I drank, the fear diminished, the anxieties and insecurities and dark suspicions I always harbored in relation to others just eased out of my bones and were replaced instead with a feeling of courage and protection. Two drinks: This isn't so hard, being out in the world. Three drinks, four: No, it's easy; it's fun. See? I can interact with people, I can laugh and talk, I can even let my guard down. Drinking was my fastest and most reliable route to intimacy, the substance that opened me up to others, gave me a voice, allowed me to share my ideas and my feelings and my body.

Of course, the courage was artificial, and the sense of protection it gave me never lasted, because it never took root in my bones. The story is old and familiar: I drank and I lost control over drinking and my life got very messy and I had to quit. Less familiar, in literature and in life, is the aftermath, what happens when you no longer have access to that bridge or the self-protection it offers. This is what happens: the fear returns, it creeps back up on you, and it leaves you struggling mightily, with varying degrees of success and self-awareness, with the question of alternatives. How to cultivate a sense of safety without drink? How to connect to others without it? How to fend off, or even merely tolerate, the range of difficult and conflicted feelings that come up in close relationships, the longings and the fears and vulnerabilities?

I suppose you could say that there, on my patio, I began to see the outlines of some answers. You take a risk, you allow yourself to feel, you don't flee. I'd sit outside with Lucille day after day, and I'd feel that torrent of emotion — joy and delight and surprise along with self-doubt and anxiety and confusion — and I'd think: This is love, pure but not simple, not at all.

3

NINETIES DOG

The puppy baffles. Lucille and I are out for a walk when suddenly, for no reason I can fathom, she simply sits down in the middle of the sidewalk. Plunk: there she is, immobile. I give her leash a gentle tug. Her neck cranes toward me, but the body is fixed, as though she's glued there.

I use my most upbeat puppy voice: "C'mon, Lucille, let's go!"

She sits and sits. I stop for a moment and regard her. From my vantage point above her, she looks hilarious: soft round puppy body; disproportionately large ears that seem to loom up above her little black face; tail curled against the sidewalk like a long, skinny piece of rope. At this age her fur abruptly changes color at the crown of her head, shifting from jet black to a soft brown, and this gives her a comical look, as though she's accidentally dipped her muzzle in paint. And yet the puppy also appears quite composed, planted there on the pavement. Despite her youth Lucille has established herself already as a creature of eminent dignity and restraint, a serious, watchful presence who always knows exactly what's going on in a room — who's there, what they're doing, whether food is involved — but never comments on it, never barks or yips or jumps unless she's provoked to do so by something that strikes her as extraordinary: a sudden noise, or the emergence of a biscuit. So there she sits, tiny and earnest, as though pondering a matter of grave import.

I tug some more, but she's like a bag of flour: dead weight, forelegs locked in front of her. "Let's go, Lucille!" At this she lies down, becoming more immobile still, and I realize I have two options: Stand there and implore, or literally drag her down the sidewalk by the neck. What is this about? So much insistence in such a tiny, fuzzy body! Is she tired? Preoccupied? What is she *thinking?*

This is the kind of minor occurrence that causes strangers to stop on the street and talk to you, briefly making the world feel like a warmer, more jovial place. They stop and they gush and coo, "Oh, a puppy! How old? How *cute!*" But I also notice something else: people are asking me the strangest questions, broaching matters I hadn't even thought to consider.

While Lucille is plastered to the pavement, refusing to budge, a woman comes up and dotes on her briefly, then asks: "Are you taking her to a puppy play group? There's a group at Radcliffe Yard, every weekend at nine A.M."

A puppy play group?

Later that same day, someone else stops and says, "Have you found a puppy kindergarten class yet? I know a really good one."

Kindergarten? For Lucille?

I stare blankly. I blink down at my dog. I feel like I've wandered into a foreign country, a place where people are speaking a different language. "Buy her cow hooves!" A man in a leather jacket corrals me on a street corner and tells me this with great authority. "They're really smelly, but dogs love to chew on them." "You need a crate." Every dog owner I talk to says this, in terms absolute and uncertain. "Gotta have a crate." Questions abound; so do opinions. What are you feeding her? Iams is best: lamb-and-rice. No, Eukanuba's best. No, Hills Science Diet. Who's your vet? Are you brushing her teeth?

Wait a minute: brushing her teeth?

When I was growing up, the dogs in our household lived simple lives and seemed to require proportionately simple treatment. They ate Alpo, purchased at the local grocery store. They hung out in the yard for much of the day, slept on the floor outside my parents' bedroom at night. They came in, went out, and save for an occasional walk around the block, they lived fairly bounded and routine lives. This is not to say the dogs weren't loved, even adored, or deeply valued, just that my parents, like most dog owners I observed, didn't seem to jump through hoops to make them happy.

Less than a decade later (my mother's last dog died in 1990), the dog owners I ran into were jumping through lots of hoops, often big expensive ones. They were enrolling their dogs in puppy socialization classes and canine summer camp and doggie day care. They were shopping at pet stores the size of Kmart, and they were trotting their dogs into all manner of places that would have been unthinkable ten years earlier — bookstores, cafés, resort hotels. They were quoting Brian Kilcommons and Carol Benjamin the way new mothers quote T. Berry Brazelton and Dr. Spock. And yes, they were getting down on their knees every evening and massaging their dogs' gums.

I was astonished by this, but I was also in the throes of puppy love, desperate to ensure this new creature's well-being, and so I plunged right in. I shopped like a madwoman in those early months: bought Lucille a little doggie dental kit, and a big fluffy bed, and an elegant set of porcelain food bowls, and a lovely fleece blanket smartly embossed with paw prints. I became obsessed with her social life, started rising at dawn to get her to a seven A.M. play group, and found myself standing around the park with other dog owners like so many moms in the playground, engaged in intense discussions about house training and feeding schedules and the proper color of stool. I acquired new gear (training collars, grooming brushes, nail clippers), and I developed new fears (kennel cough, whipworm), and I became preoccupied with new questions (kibble or canned food? regular or retractable lead?), and every once in a while, when I'd catch myself poring over a copy of *Dog Fancy* magazine at the pet shop instead of sitting at home with *The New Yorker*, or examining the protein content on the back of a bag of dog food instead of making my own dinner, or wanting to curl up in bed with the puppy instead of my boyfriend, I'd look up and think: What happened to my life? Have I gone completely mad?

The facile answer is: Yes. Absolutely. I am precisely the kind of dog owner the media likes to make fun of: over the top and heading for the deep end. You see evidence of this everywhere, snappy little stories about indulgence and excess. *USA Today* writes about people who book their pets rooms in hotels that offer bone-shaped beds and special doggie room-service menus; *The New York Times* reports on Manhattan dog owners who are spending up to $10,000 per year on their dogs, springing for everything from catered birthday parties to canine lingerie; *People* magazine gushes over a health-and-fitness center for dogs in Westwood, California, that features treadmills, Jacuzzis, and swimming pools specifically designed for dogs. The message: those wacky Americans, gone from pet rocks to pet dogs.

The real answer, of course, is more complicated; it has to do not with American obsessionality or faddism or caprice but with need, and with the emotional niches a lot of modern-day dog owners are asking their animals to occupy. Not long ago I had a long conversation on the phone with a woman named Vicky, a forty-one-year-old photographer who lives in Manhattan with her dog, a two-year-old chocolate Lab named Thurston, and we got to talking about how much the world of people and dogs has changed since we were growing up.

Our talk was peppered with precisely the kinds of silly, indulgent details you read about in the press (both our dogs have celebrated their birthdays with parties in which Frosty Paws, a canine version of ice cream cups, are served; both our dogs like to sleep under the covers; both our dogs attend regular play groups and have frequent dog "dates" with their best friends; and no, our parents' dogs never enjoyed perks like this), but the underlying content was more serious: at heart we were talking about social change and the urbanization of America, about a new reality in which the emotional roles dogs play have been thrown into increasingly stark relief.

Vicky grew up the way I did, in the era of the yard dog. "We didn't have dogs hanging around inside with us all day," she says. "They pretty much lived outside. My mom let the dogs out in the morning, and they hung out in the yard all day, or someone left a gate open and they went roaming around the neighborhood until dinnertime. Sometimes the cops would call and say, 'Hey, we've got your dog,' and he'd be, like, in another *town*. We didn't have leash laws. We didn't really have to think about exercising the dog. We didn't have to think about *training* the dog, for that matter. The dogs were just — you know, *there*. They just weren't in our faces the way they are today."

Thurston, by contrast, is very much in Vicky's face. She lives in a small one-bedroom apartment, on the fifth floor, and Thurston is a big, noisy dog: she is far more conscious of his presence, his needs, the quality of his life and of their relationship. If he doesn't get exercise, he's at her constantly: nudging her, trotting up and shaking a toy at her, barking. If the dog needs to go out, he has to ask, and she has to bring him outside, on a leash. If she hadn't taken the dog to obedience classes, her life would be a living hell. "Imagine living with a seventy-pound Lab who doesn't understand the words 'stay' or 'off' or 'quiet,' " she says. "I'd be out of my mind."

Thurston is also a big emotional presence in Vicky's life, as well as a physical presence, and she says this strikes her as a radical change from the past, too: it's the part of our conversation that has to do with social change. Like me — and like one-quarter of the U.S. population — Vicky lives alone, without the emotional or financial support of family, and she often feels isolated, prone to fits of loneliness. Like me (like all of us), Vicky also lives in a time of transience and instability, in a culture in which one in two marriages end in divorce and 21 million women are divorced or single mothers, and this has left her with a kind of big-picture confusion about relationships: Does she want to get married? Not sure. Have kids? Doesn't know. Continue to live alone? Maybe; who knows? We talked for a while about our mutual sense of rootlessness — about living in an environment where friends relocate to new cities and new jobs every other year, where you yourself might change apartments six times in a decade, where you barely recognize your neighbors when you pass them on the street — and then we talked about the extent to which dogs can offset this feeling, about how profoundly stabilizing their presence can feel. As if to illustrate the point, Vicky stopped midsentence and asked me, "Where's Lucille right now?"

I looked down: dog at my feet, lying flat on her side, fast asleep. "Right here," I said. "Sleeping."

"Same here," she said. "Big old Lab, snoring on the floor."

And there you have it: in the midst of uncertainty, there is the dog, one sure thing, stability in fur.

The personal voids that dominated my landscape when I ventured out to the shelter to find Lucille are in many ways cultural voids as well, ones that have been blasted open by thirty years of social upheaval. Loneliness. Transience. The breakdown of family and the search for alternative sources of support. The stresses of life in urban America, which is at once more crowded and more isolated. And, in the midst of that, 55 million pet dogs. As indulgent and fanatical as the modern dog owner can seem (even to *me*, as I'm doling out the Frosty Paws), I understand the impulses behind such behavior: you give a lot because you have ended up with a creature who seems to give back so much in return, who's literally thrilled to see you every single time you walk in the door, who's always in a good mood, who's always *there*. "I'd do anything for my dog," Vicky says. "Thurston is like . . . well, for lack of a better term, he's like my *boyfriend*. He's there on the sofa when I watch a movie. He's there on the bed when I go to sleep. He's who I hold when I've had a bad day, when I cry. It's really very intimate, and I think that's something dog people just didn't experience when I was growing up."

My paternal grandfather, a wealthy man who fancied himself a gentleman of leisure, would have agreed. He lived in upstate New York, farm country, and he always had a passel of dogs around: hounds, mostly, who'd accompany him on hunts. He liked dogs — admired them, valued their skills — but he certainly didn't dote on them, or depend on them emotionally, and he would have been appalled by the kind of life I lead with Lucille: dog in the bed? An outrage. His dogs rarely made it through the front door.

My father would have had a more charitable view of my relationship with Lucille, but I'm not sure he would have emulated it with a dog of his own. Dogs are dogs, he might have said: nice family pets, highly amusing creatures, and clearly objects of attachment, but not primary relationships. His perspective lay somewhere between my grandfather's and mine: sure, the dogs can come inside, but please, keep them off the furniture.

And then there's me, last stop on the continuum. Photos of the dog on every available surface of my home. Piles of dog magazines and catalogues from pet supply houses stacked on tables. And, of course, the dog herself, snuggled wherever she damn well pleases: on the sofa, on the bed, on the leather mission chair.

Together the three of us tell a story about people and dogs, about the contract that's bound us over the years and how it's changed. In my grandfather's day (and for generations before him), people acquired dogs because they wanted them to execute specific tasks — herding sheep and cattle, guarding property, leading hunts. Their arrangements with dogs were rather businesslike in nature, utilitarian, and if my grandfather's dogs occupied any emotional role in his life, that would have been a fringe benefit, the bonus of acquisition but not the primary motivation. By the time my parents bought their first dog, in the early 1960s, suburban America was in full flower, and the majority of dog owners no longer needed dogs to herd or hunt; the working dog gradually mutated into the suburban dog, acquired as an adjunct to family life and subject to relational rather than utilitarian terms of service. My parents got dogs as companions, as playmates for the kids, and they chose elkhounds not because they required partners to lead them on elk hunts through the streets of Cambridge but because they liked the look of the breed, its intelligence and temperament. And me? If Lucille and I had an official contract, the concept of work wouldn't even make it into the fine print. As I write this, she's under my desk, lying on a soft fleece bed, and I consider this to be a key part of her household duties: keep me company while I tap away at the computer; be my ally, the steady presence in my life; give me something to take care of, to touch, and to love.

Intimacy has always been a feature of the relationship between people and dogs — without some capacity to bond with and nurture dependent pups, humans never would have let the dog's early ancestors into their homes, let alone their hearts. But at least in this country, emotional closeness has never been such an essential part of the bond, and it's never been so overtly expressed. In one of the first dog play groups I attended, Lucille's companions were named Sadie, Max, Franny, Murray, and Marty; we owners used to joke that, together, the dogs sounded like members of a weekend tour group in the Catskill Mountains, but I think we also understood, without talking about it, that the choice of human names reflected this heightened emotionality. Easily half of today's dog owners name their dogs after people; the majority of today's dogs are allowed to sleep in their owners' bedrooms, almost half of them in bed with a family member; intimacy for people like me is less a by-product of acquiring a dog than a goal, its raison d'être.

A woman named Kathy — single, a schoolteacher, late thirties — lives with a Wheaten terrier named Guinness. When I ask her why she has a dog, she says simply, "Dogs are our children." I nod. I wouldn't draw that exact parallel, but I know what she means: Lucille is a primary focus of caretaking for me, the creature I nurture. A man named Donald, who lives with three dachshunds, tells me, "I get a lot of my physical contact and affection from my dogs. If you don't have a wife and kids and you're not the kind of person who runs around hugging your friends all the time, the dogs are *it*." I nod again: I probably kiss Lucille forty times a day, reach down and just touch her more times than I can count, which is not something I can say about the humans in my life. Paula, who is painfully shy and troubled and thin, looks down at her hands when she talks to me about her standard poodle, Bridgette. "The dog is family to me," she says. "I'm much closer to her than I am to my family." Paula came from a household that emphasized appearances: her role was to be a well-behaved, good little girl, someone who would reflect well on her socially conscious parents, and she grew up feeling undernurtured, confused, isolated. Bridgette gives her the opposite feelings, a sense of being loved, sure of herself, connected. "She's the reason I belong in the world," Paula says. "Otherwise I'd be lost. I wouldn't know who I was. She's how I define myself." This rings many bells for me, reminds me of how vague and shapeless my life felt when I first came across Lucille, so I nod yet again. Dog as family member. Dog as primary source of emotional support and affection. Dog as object of self-definition. We may have appreciated the dog's ability to play those varied roles for centuries, but until recently many of us have not needed him to do so with quite so much intensity.

This is not to say that all modern dog owners bring high degrees of psychological complexity to their relationships with dogs. But many of us are living in closer emotional proximity to our dogs than we might have in simpler or more stable times, and sometimes, when I contemplate my attachment to Lucille — her centrality to my life — I think about that difference, about how fundamentally different our world is from the one my parents and their dogs enjoyed. Toby and Tom, the elkhounds, lived before the era of the dual-career couple, and they had a whole family at their disposal, five humans to attend to their needs at various points during the day. My mother worked at home, so she kept the dogs company in the morning; the kids would come home after school and play with them. (I remember spending an entire afternoon when I was about twelve "teaching" Tom to hunt down Fig Newtons, which I hid in various spots around the living room; he was a quick study, consumed an entire box, then threw up all over one of my mother's oriental rugs.) The dogs might have lived simple, bounded lives, but they were never lonely. By contrast, I am *it* for Lucille. I work at home, so I don't have to contend with the guilt of leaving her from nine to five, but I am also the only human in her life, the only person who walks her or feeds her or gives her something to watch, and I'm often struck by the sense of responsibility this gives me: how much at my mercy she is.

My parents' dogs also lived in a safer world than Lucille and I do. When Toby escaped from the yard, which he did with some frequency, my mother worried about him, but the streets were less congested than they are today, and the drill became fairly predictable: an hour or two would pass, the phone would ring, an annoyed voice on the line from Dunkin' Donuts would sigh and say, "Mrs. Knapp, would you please come and pick up your dog? He's scaring the customers." Me? If Lucille got out of the house, I'd worry about everything from city traffic to local lunatics: someone might run her over, haul her off to animal control, steal her and sell her for medical research. Unless I choose to take her out, Lucille is an indoor dog, and this means she is around me more, witness to more intimate aspects of my daily life: she's staring at me intently every time I reach into the refrigerator; she's poking her head into the bathroom while I shower; if I'm having an angry conversation on the phone, she's skulking nervously out of the room; were I to bring a new man into my life, she'd be there on the bed with us (and God only knows how she'd react) while we made love.

Dog owners like me are in closer emotional proximity to dogs these days because we understand them better, too, thanks in large part to the explosion of information about the nature of the dog, his heritage and his mind. My grandfather certainly didn't spend his leisure time cruising dog chat rooms on the Internet, or investigating canine Web pages, or attending on-line veterinary forums. Nor did he stand around in bookstores poring over books about pack structure and canine intelligence and the emotional lives of dogs: books like that didn't exist in the mass market. By contrast, the first month I had Lucille, I soaked up dog books. I read Barbara Woodhouse (*No Bad Dogs*) and the Monks of New Skete (*How to Be Your Dog's Best Friend*) and Brian Kilcommons (*Good Owners, Great Dogs*), and I remember the seriousness and emotionality with which all three treated the bond. The message, consistent in each: You are the pack leader; you are dealing with a sentient being; acquiring a dog is a relational undertaking, not a proprietary one. Kilcommons's books says as much in the subtitle: "The Kilcommons way to a perfect relationship." The bond — its proper cultivation — is at the heart of the Monks' methods as well: "We approach training," they write, "as a way of relating to your dog," a concept that Woodhouse casts in an even more emotional light. "In a dog's mind," she writes, "a master or a mistress to love, honor, and obey is an absolute necessity. The love is dormant in the dog until brought into full bloom by an understanding owner."

. There's the new contract in a nutshell: love. I will love you; you will love me.

★ ★ ★

It gives me a little chill to think about what Lucille's life might have been like had she grown up with a man like my grandfather. No soft bed or fleecy blanket, no jaunts to the dog park to see her friends, nary a Frosty Paw. I don't believe that dogs really *need* all those canine frills, but I get a great deal of pleasure out of indulging her, and I believe that, at heart, Lucille enjoys our shared intimacy as much as I do. Every once in a while I'll hear a thump, thump from under my desk, look down, and realize she's wagging her tail in her sleep; she's probably got squirrels on her mind, but I like to think that maybe — just maybe — she's dreaming about me.

And yet I'm also aware that this new, more intimate contract, with all its complicated emotional riders, has its trade-offs. Once, when Lucille was about a year and a half old, I took her lure-coursing, an activity that takes place in a big, open field and basically allows dogs to go haywire with their predatory instincts: a white plastic bag is mechanically swept along a track at high speed, and the dog, seeing what must look like a rabbit in flight, gets to chase it down. Lucille went bananas. We got to the field, and she saw that plastic bag zipping back and forth along the course, and her whole body tensed in excitement. She let out a sound I'd never heard from her before — a shrill, high keening — and she strained at her leash until she nearly choked, and when I finally let her go, she tore across that field with an energy that seemed almost feral, a wildness I found glorious to behold. I watched her run, saw the concentration and drive in her bearing, and I thought about the discrepancy between what she was bred for (herding, chasing) and what's expected of her (lounging quietly under my desk, sleeping with me at night). Make no mistake: I know Lucille has a great life — lots of contemporary dogs have great lives — but I'm also aware that she lives in a modern, urban, human home, an environment that's designed to stanch many of her natural instincts and abilities. Be quiet. Off. Stay. *Leave it.* The underlying messages behind those commands are often none too subtle: conform to *my* world; respond to *my* needs.

Leslie Nelson, a Connecticut-based trainer, uses the phrase "the Lassie syndrome" to talk about the gap between who dogs are and who we expect them to be, a term I've heard echoed repeatedly. Lassie was both a product and an emblem of suburban America, and as such, she both represented and helped seal the dog's transformation from working partner to family companion. By the mid-1950s, televisions were installed in the majority of American homes, and Lassie was bounding across all of them, teaching an entire generation of dog owners what to expect from their four-legged family members: loyalty, selflessness, and above all, perfect obedience.

In a sense Lassie underwrote our new, more relational contract with dogs, and trainers and behaviorists are nearly unanimous in their opinion that she did more to shape (if not warp) contemporary American fantasies about dogs than any other cultural icon. "A lot of people think dogs were put on earth to live in suburbia while we're off working, and then to lie by the fire at night at our feet," says Nelson. "It's Lassie. It's good old Fido. We come into owning a dog with a set of expectations that are incredibly unrealistic. The truth is that very few dogs were created to be household pets: they have needs that living in a pet home just doesn't fulfill."

Owners themselves are often painfully aware of this disparity. "Sometimes I feel like we spend an awful lot of time trying to keep the dog from being a dog," says Sara, who lives outside of Boston with her husband and a two-year-old border collie mix named Elmore. Elmore is a rather delicately built and high-energy dog — forty pounds, with a wispy black coat and the intense, eternally fascinated gaze of a border collie — and he is by all accounts a perfectly happy animal. Sara's living room floor is littered with his toys — rubber balls, rope tug toys, rawhide bones — and the dog spends a great deal of time outdoors on a run, where he races back and forth barking madly at squirrels, passing cars, and (especially) UPS trucks. But Sara looks around and sees an environment of trade-off and compromise. There, in the corner of the room, is Elmore's crate, where he tends to hang out for the five to seven hours he's left alone each day: guilt city to Sara, despite her belief that Elmore feels safe and perfectly content when he's in it. There, on the coffee table, is a pamphlet from a company that designs and installs electronic fences — Sara is debating whether he'd be happier if he could run free in the yard, without being tied to his run, or whether the periodic electric jolt would constitute cruelty. "Sometimes I think he has a great life," Sara says. "I mean, we take this dog to McDonald's and get him his own order of Chicken McNuggets. But other times I look at him and think, 'God, I wish I had a hundred-acre farm and a flock of sheep.' Wouldn't he be happier if he had that kind of life?"

Vicky struggles with a more subtle issue. "I don't really worry about Thurston not having work to do — he gets a ton of exercise — but I sometimes worry if it's healthy to be this close, for either of us." She worries about mutual dependence, an excess of it. Thurston won't let her out of his sight for a minute; she steps out of the shower and finds him lying on the bathroom rug waiting for her, and she wonders if this is a bad sign. She worries about his exquisite sensitivity; she can't so much as raise her voice without upsetting the dog. If she needs to cry, she'll often shut herself in another room so she won't have to see Thurston's eyes, that pained, empathic expression gazing back. She even worries about how much she worries about him. "I hate that 'Oh, he's just a dog' mentality," Vicky says, "but sometimes I wish I had a little more of it in me. Sometimes I wish I was a little less intense about him. Do you know what I mean?"

Indeed I do. Intensity with a dog is a complex thing. In our other key relationships — with family members, lovers, close friends, therapists — we can sit down and talk about our feelings: what we need and want, what's frustrating or disappointing us, where the relationships stands and where it's going. This, of course, is impossible with the dog, and the gulf between us can be utterly baffling, by turns a source of joy and one of enormous frustration. Sara wishes she could sit down and *ask* the dog: Do you really like your crate? Do you know we're coming back when we leave you alone? Would you rather be in here or out there on your run? Vicky wishes she could "get inside Thurston's mind for ten minutes and find out if he's really happy, if he worries about me too much, what's going on in there."

These are the kinds of issues that crop up when you live in close relationship with an animal. You may get the best of what dogs have to offer — their capacity for companionship and loyalty, their ability to make us feel connected and needed. But the business of feeling intimate with a dog, considering him or her to be a part of a vital relationship, can be confusing and strange, too. You battle, sometimes daily, with communication, or the lack of it. You struggle with guilt, and with the unknown. And sometimes, through the hazy, human fog of your own attachment, it can be very easy to forget that the dog is the dog.

This happened to me all the time those

first few months. Lucille and I would be driving along in the car, and stopped at a traffic light, I'd hear myself saying something to her, something very benign like, "Hey, you: how's it going back there?" And then I'd turn around and look at her, and I'd see this little *puppy face,* jet black, those little triangle ears flopped over at the tips, and I'd actually feel a little jolt of surprise, as if I'd suddenly realized: Oh, my God, there's a animal in my car!

I'd search her face: "You have no idea what I'm talking about, do you?" The puppy would stare back at me, curious and alert and undoubtedly baffled, no idea at all. And I'd just shake my head. What a strange sensation, to look down and remember that you're talking and interacting with an animal, a member of a different species: it drives home their otherness. The dog is not a creature who experiences communication and connection the same way I do. She is not a being with access to language or human constructs, and she is not a perfectly attuned, cleverly disguised version of a person in the backseat with a clear, knowable, or even remotely human agenda. The dog is, in fact, the dog.

4

BAD DOG

The dog is rolling in decomposed squirrel. We are at Fresh Pond, a local reservoir that's ringed by a two-mile walking trail, and she has disappeared up a wooded hill and into the brush. I can just make out her silhouette through the trees, that unmistakable slow-motion dip: she lowers her shoulder over the carcass, eases her body down and to one side, and then casts herself upon the rot, throws her whole body into it, squirms against it until her neck is encrusted with squirrel innards, stiff and rank and (to her) utterly delightful.

Ah, Lucille. As she does so well and so frequently, the dog is reminding me that she is very much a dog, and at the moment I am finding this deeply annoying.

So I call her.

"Lucille, come!"

Nothing happens.

I try again, a little louder.

"Lucille! *Come!*"

Nothing. No dog.

My voice grows louder, more urgent.

"Lucille! Come! Now!"

This time I see her stand up, cast a fleeting glance my way, then drop down onto the squirrel a second time, hind legs flailing in glee. My irritation mutates into anger, a feeling that's tinged with an odd sense of betrayal. Lucille is a year old by this point and she knows better. She knows what "come" means. She is totally dismissing me, and I can't help but take it personally. I look around, hope no one is observing this obvious failure on my part (Who's the moron who can't control her dog?), and this time I scream it: "LUCILLE! GET OVER HERE, *NOW!*"

Finally Lucille comes trotting out of the brush and gives me a look of utter innocence, the perfectly merry pup. I know I am not supposed to yell at her when she finally shows up. I understand that the act of coming to me — even belatedly or reluctantly — should never be associated with punishment, but at that moment I am so angry, I want to spit.

My frustration is a blend of irritation and fear: it is, quite simply, a pain in the butt to have to stand there and scream into the bushes for your dog; it's also very scary. The trail we're on abuts a fairly major roadway, and when she disappears into the woods that separate the two, I often have no idea if she's up there rolling in something putrid or if she's strayed farther, chased a live squirrel out of the woods and into the street. But the sense of betrayal is more complex, so powerful at times it takes me by surprise. When the dog fails to come on command, when she ignores me like this, I feel some well of fear rise up about being inadequate, unworthy of attention, out of control. I hear a small voice inside: She doesn't come when you call, because she knows you're a wimp. She doesn't come because she doesn't love you. You're a loser; you can't even control your dog.

Human emotion, meet the dog. Fantasy and ego, meet instinct. Lucille is for the most part a wonderfully responsive animal — intelligent, capable, actually voted "most likely to succeed" in her puppy kindergarten class (and I have the diploma to prove it, dutifully taped to my refrigerator) — but she is also a dog, which means she is governed by drives and instincts that often have little to do with me. And so, on occasion, we clash. She acts in ways that run counter to my vision of an ideal dog; my feelings get hurt. Together, we enter the murky, emotionally loaded continent that every dog owner must visit, a land called Obedience and Control.

On the surface the landscape here looks pretty straightforward, eminently navigable. You read a few training books, maybe enroll the dog in an obedience class, get the basic rap. You learn that dogs are creatures of hierarchy and order who need to know their place in the pack. You understand that in the wild the pack leader or "alpha dog" makes up the rules and enforces them, decides when the lower members in the pecking order eat, determines who can get away with what, settles conflicts. You recognize that as a dog owner, this is your job as well. Be the alpha. Control that dog.

But this turns out to be a bit harder than it sounds, in part because dogs can be difficult to control (they can be stubborn and willful, fast and strong, assertive and pushy), in part because humans, at least some, have a tough time acting in ways that communicate authority (directly with clarity and resolve), and in part because in the act of controlling a dog (or trying to), you stumble upon one of the central differences between us and them: dogs live in a physical world, we live in an emotional one. Their universe is about immediate drives, smells, sounds, pleasure, pain. Ours is about feelings, fantasies, symbols, abstract thought. They act, we interpret, and in the gap between those two modes of being, there is a whole lot of room for confusion and conflict.

Consider the recall, which is probably the single most important command in an owner's repertoire and the one that gives owners the most trouble. When Lucille was a young puppy, I liked the idea of taking her to Fresh Pond and letting her run off-leash. I liked the fantasy behind it: she'd run and cavort, I'd amble along, and periodically she'd race back and walk by my side, gazing up at me in love-struck devotion. This is such a common and powerful fantasy about the human-dog relationship it's hard to see what it really represents: it's an image of human power and canine subservience, ready-made and non-negotiable. It places you at the center of your dog's universe, and it speaks to our most central ideals about canine love and respect: that those qualities are unwavering and automatic, requiring little to no effort to attain. Dogs stay with us (and come back to us when we call them) because they love us and want to be with us.

This fantasy is complicated at least in part because it seems relatively easy to realize. At some of the dog groups I went to with Lucille early on, I'd see owners saunter in and out of the park with their dogs off-lead, the dogs trotting calmly by their sides, and I'd feel a little jolt of longing, an Oh-I-want-that sensation. The image got tied up (and quickly) with all those deeper longings about connection and family: I might not have parents, I might feel uncertain about the other humans in my life, but I'd have this *dog*, this adoring dog who'd be glued to my side, in my thrall, the very picture of attachment. So day after day when she was a puppy, I'd go to Fresh Pond with Lucille and I'd unclip her leash — and she'd promptly disappear into the brush, compelled not by any complementary fantasies about attachment and connection but by her nose.

My own behavior, of course, flew in the face of everything I'd read, everything I'd been told: never call the puppy unless you can reinforce the command; never call the puppy unless she's at the other end of a leash and you can reel her in if she doesn't come *immediately;* if you can't back up the directive, you're allowing her to decide whether or not she feels like coming, and nine times out of ten, she will decide that there's something far more interesting out there than you. Which is precisely what happened. Lucille would lag behind, and she'd stop every fifteen seconds to sniff something, and she'd meander off into the trees, and I'd find myself turning around every twenty-five feet and imploring her to keep up. Sometimes she would trot up to me immediately when I called; sometimes she wouldn't; within the space of about a week I'd made the huge, not-quite-irrevocable and very common mistake of teaching her that "come" was an optional command, and the word lost its meaning entirely. I'd scream, I'd shout, sometimes I'd get so frustrated, I'd sit down in the middle of the path and cry. And Lucille? She'd amble my way if and when she wanted to.

Like I said, human emotion, meet the dog.

And yet control is such an individual matter, such a complex blending of owner and dog and the traits they both bring to the bond. Breeding and temperament are the big wild cards here, defining control issues in very specific ways: get a Samoyed, bred as a sled dog, and you will spend a lot of time lurching along at the end of the leash shouting, "No pull!" Get an Australian shepherd, and you will spend months (possibly years) trying to teach the dog not to herd joggers. Terriers dig, Labrador retrievers hurl themselves into bodies of water, scent hounds career across fields, and these instinctual drives will be muted or highlighted by a dog's nature. Some dogs are more dominant than others, some more placid or high-strung, some more trainable, some (frankly speaking) more obnoxious, and all of these qualities can shape your relationship with control (and with your dog) dramatically.

Our skirmishes at Fresh Pond aside, Lucille tends naturally to limit my battles on the control front, evoking pride more often than frustration and insecurity. As a puppy, she took to training quickly, learned the commands "sit" and "down" within a matter of minutes, and was housebroken within days. She still seems to enjoy obedience work today, knowing what's expected of her and then carrying out the task: she can execute a mean high-five and a perfect high-ten, she will hug me upon request, and when we practice heeling off-leash, she trots beside me with her ears flattened in pride, tail swishing behind her as if to say, "I can do this. See how good I am?" Lucille is also an exceptionally mellow dog, perfectly content to curl up in whatever room I happen to be in and sleep. The second or third time she met Lucille, my house cleaner asked me how old she was. "Eighteen months," I said, and her jaw dropped. "You're kidding! I thought she was, like, *eleven*." The dog's temperament affects the texture of my life significantly, and I feel blessed by this, not just because I appreciate her manner (which I do) but because I know how stressful I would find living with a more assertive, noisy, or high-strung dog.

But to each his own. Human needs for discipline and order are all over the map, and a dog's behavior might bring out the control freak in one owner, the wimp in another, the casual pal in a third. Just the other day, at Fresh Pond, a runner came jogging toward me and called out, "Hey, have you seen a big golden retriever?" I said no, and he shrugged as he ran past me, smiling. "Oh, well," he said, "he'll find me." The man was completely unfazed by his dog's absence, possibly because he knows from experience that his dog *will* find him, possibly because, unlike me, he's not the sort of person who gets bogged down by questions about an animal's love and respect, let alone safety.

Not ten minutes later, a woman approached, someone I periodically see walking a Pomeranian. The dog, a little rust-colored fuzzball, is always on-leash, and the woman tenses visibly when she sees other dogs coming toward her: she pulls the dog in closer to her side, then picks the animal up until the approaching dog is safely past her. Sometimes I hear her whisper little reassurances to the dog — "It's okay, honey" — and a few times she's smiled shyly at me and confessed in a hushed voice: "She's afraid of other dogs." In fact, there is no physical indication that the dog is afraid — her hackles aren't raised, she doesn't bark or snarl or seek protection from her owner when she sees a dog approach — but no matter: one owner's nonchalance is another's perpetual state of terror.

And one owner's definition of control is another's recipe for chaos. Alison, an owner from San Francisco, tells me over the phone that she considers a quiet, controlled demeanor to be part of her dog's "job description," and she has taught the dog, a sheltie named Sky, to lie in a silent down-stay whenever she is working at her computer or talking on the phone. Sky is not allowed on the furniture, never begs and never barks indoors, having learned the command "quiet!" very early on.

Alison's summation of Sky: "She's a *great* dog."

Janet, an ER nurse from Boston, would find this laughable, unthinkable. She lives with a fourteen-pound Pomeranian-terrier

mix named Kim who appears to be constitu-
tionally incapable of sitting still: the dog
yips constantly, runs nonstop laps around
the living room, leaps up and down as
though attached to permanent springs, and
begs for biscuits constantly. Janet, who
admits she "thrives on chaos" herself,
adores this about the dog.

Her summation of Kim: "She's a *great*
dog."

Breed and temperament of dog, personality and style of human, tolerance for
chaos, feelings about authority and definitions of love: control (or lack of it) is what
happens when you throw all those variables
into the great stew pot of contemporary human-dog bonds, add a pinch of ego or a
dash of insecurity, and let them simmer.

Want to see how many different dishes
you can come up with? Just stand around a
dog park and watch how owners react when
their dogs start engaging in that less-than-
endearing canine behavior, crotch-sniffing.
I have experienced the full gamut of responses: some strange dog pokes his nose in
my crotch, and his owner leaps up, aghast,
and reprimands the animal: "Stop it! Bad
dog!" Another dog does the same thing, and
his owner stands idly by, pretends not to
notice, perhaps even gets a secret kick out of

the transgression. A third does it, and the owner reprimands *me:* "Oh, you must have treats in your pocket." (Translation: It's my fault.) What's fascinating about control is not just that we have to decide whether and how wield it; it's what those decisions have to say about us, about who we are and how we experience relationships.

My first year with Lucille, at a park I no longer go to, I used to watch a woman named June with her puppy, a Scottie named MacGregor. MacGregor was about five months old, a fierce little male and very headstrong, as Scotties are wont to be. June adored him — she used to swoop into the park her first weeks with him and announce, "I am in love with this dog! *In love!*" — but as time wore on, she had an increasingly difficult time with him: MacGregor wouldn't listen to a word she said, wouldn't come when she called, wouldn't sit when she commanded him to, tested her constantly. On the face of it, this seemed odd, because June was very fierce and headstrong herself: she wasn't the sort of person who gets wimpy or fearful in front of a dominant dog, who just stands there and shrugs her shoulders. She'd give very clear commands — "MacGregor, *down!*" — and she didn't let him get away with disobeying; if he didn't lie down right away, she'd *make* him. "I don't know what to do," she'd complain, and as the weeks wore on her complaints intensified. MacGregor had begun snarling at her in the house, baring his teeth and growling when she tried to take away a toy; twice, as she tried to wrestle him into the bathtub, he bit her on the wrist.

Books on obedience training would no doubt call this a classic case of dominance and aggression: the dog is vying for top position in the household, and the owner needs to assert her authority in more direct and deliberate ways. Firmer corrections. An end to any household routines that might communicate equal status to the dog, such as allowing him to sleep on the bed. But the longer I watched MacGregor and June, the clearer it seemed that she participated in this dance of dominance with him: the dog's behavior appeared to touch her on a primal level, stirring up a range of ancient emotions that caused her to react to him in ways that fueled his dominant streak.

June was a fairly self-disclosive woman, and she talked a fair amount over the weeks about her mother, who apparently had a strong aggressive streak herself: a "controlling bitch," June called her, "Attila the

Mom." June herself left home at age eighteen, after a lengthy rebellion-filled adolescence, and did her best to keep contact with her mother to a minimum. What she did less effectively, it seemed, was keep her feelings about this battle-ax of a mom in check. One day I watched her put MacGregor in a sit, then set a biscuit in front of him and tell him to stay. MacGregor sat for a few seconds, then stood up and made a move toward the biscuit. I watched June. She lunged toward him. She screamed an ear-piercing *"No!"* and then she picked him up by the scruff of his neck, screamed *"No"* again, and literally hurled him back down onto the grass. Her hands were shaking she was so mad, her face contorted with rage. She looked as if she was about to go after the puppy again when an onlooker, apparently horrified, suggested to June that her reaction may have been a bit out of proportion. "Dogs have excellent hearing, you know," she said. "You really don't have to yell at them so loudly." June looked a bit taken aback, then tried to laugh the incident off. "Oh, my God," she said. "I'm turning into my mother."

Having watched and listened to this woman for a while, I suspect there's truth in that. June felt helpless in front of the dog's assertiveness. She felt enraged by it. It seemed to touch some of her deepest feelings about dominance and control, about both wielding power and losing it. And while the dog may indeed have been a naturally dominant, willful little pain in the butt, it seemed equally plausible that her reaction — testing, confrontational, angry — contributed to the conflict. He was rebelling, just as clearly as she had as a teenager. They'd entered into a classic power struggle, a grown woman and a five-month-old puppy.

And yet this is exactly what amazes me, the way a person's core sense of self — anxieties, insecurities, grandiosity, fantasies of self and other — can come bubbling up when it comes to controlling a dog, even in the smallest and seemingly most inconsequential ways.

At the same park where I used to watch June and MacGregor, a woman named Ellen used to stand on the sidelines and deliver a running commentary about other people's dogs. Ellen, a grad student in her mid-thirties, was opinionated and rather brash, a tall heavyset woman who secretly annoyed me by consistently referring to Lucille as "Lucy." She'd watch a dog beg someone for a biscuit, and she'd huff, "I would *never* let my dog beg." She'd see an owner trying to call her dog, watch the dog dart away as soon as the owner got within reach, and she'd say, "My dog would *never* do that." In fact, Ellen's dog, a three-year-old chocolate Lab, appeared to be no more or less well behaved than anyone else's dog, but Ellen was so invested in the idea of her own power that that didn't seem to matter. She might as well have been wearing a T-shirt: "I Have Total Control." Needless to say, Ellen was one of those owners who waltzed in and out of the park with her dog off-lead, and this grated on me, not because it implied that she did, in fact, have total control over the dog but because it sparked a competitive streak in me, tapping in to those gnawing insecurities about Lucille's attachment.

One afternoon at the park I called Lucille toward me, then gave her a biscuit when she complied. Ellen looked over at me and said, "Oooh, you're cheating!" Small enough incident, but it revealed volumes about the politics of food treats, about how something as simple as handing a dog a biscuit can become imbued with deeper meaning. Her view (not uncommon) has to do with attachment: the dog should come because she wants to, not because there's a biscuit at the other end; the dog should obey because she loves me, because she has an innate desire to please me. Actually, I've come to believe that that's an irrational view, or at least a naive one, and I tend to side with the Monks of New Skete on the question of the canine desire to please: dogs, they say, care a lot less about pleasing humans than they care about pleasing themselves; if acting in a way that pleases you means something good will happen to them — they'll get a biscuit, a reward, a pat on the back — they're likely to be motivated to carry out the task, but their agenda is not necessarily driven by the pure and selfless wish to make *you* happy. But I also know how tempting it is to equate obedience — and the control it symbolizes — with love, and I remember the little flash of worry and shame I felt at Ellen's comments: Uh-oh, maybe there's something amiss here. Is the dog coming when I whistle because she's attached to me, or is this all about liver? Am I nothing more to her than a biscuit vending machine?

Good question: I still ask it periodically. I love the dog in a very human way, which is to say that I often nurture and tend to her in the way I might nurture or tend to a person — say, a young child. I cuddle with her on the bed. I walk into a room and see her lying somewhere — the couch, her dog bed — and I go over to her and scritch her around the ears, pet and coo at her simply because I find her irresistible. And sometimes — when I'm handing her a biscuit just because I feel like it, or when I'm enthusing over her just for being cute — I look into her little dark face and I wonder: What do you *make* of this? Do all these outpourings of emotion make you feel warm and cozy and attached to me, or do you think I'm a nut?

"Sometimes I think Lucille loves me, and sometimes I think she sees me as the butler." I said this to a friend not long ago, and we laughed about it ("Time for my walk, James"; "More liver, James"), but the statement actually reflected the presence of a low-level tension on my part, a quiet but persistent worry about the relationship between power and forcefulness (mine) and respect (hers). When I indulge Lucille, stroke and gush at her, does she read my behavior as loving, or does it make her see me as a wimp? Am I, in fact, the leader in her eyes or the two-legged servant, a slightly strange but ultimately benign presence at the other end of the leash?

These questions reflect older, more personal concerns on my part — historically I've not tended to see myself as a particularly powerful or authoritative person, and I worry that the dog picks up on this, that she very adeptly sniffs out my wimpy side — but they're also valid worries and not uncommon ones among owners. The dog, after all, is a creature who by nature requires leadership and limits, who does best when she knows her place in the hierarchy. Lucille may not be an inherently dominant dog, and she may not be particularly interested in challenging my authority, but I sometimes wonder if I stray too far across some line that's important to dogs, veering out of the realm of Leadership and into the land of Love.

There are various schools of thought on the relationship between human forcefulness and canine respect, but the world of obedience trainers can be roughly divided into three camps. The first camp is harsh and unforgiving, and it is expressed most vividly by the late Bill Koehler, a legendary California-based trainer (he made a successful career by, among other things, training dogs for Walt Disney movies) who would have looked at someone like me with utter contempt. Koehler was crusty and hard-boiled in approach, and his training books are merciless on the subject of over-emotional owners, whom he refers to variously as "wincers," "humanaics," and "cookie people." "Their common calls," he once wrote, "are: 'I couldn't-do-that — I couldn't-do-that,' and 'Oh myyy — oh myyy.' " I'd be a classic wet noodle in his eyes, the kind of spineless dog owner who tries to lure her puppy around with biscuits and has no idea how crafty, uppity, even disdainful a dog can be. Koehler believed that control needs to be unyielding and absolute, discipline applied in the firmest, most unequivocal and physical ways. Without it, he believed, dogs lose respect for owners (and they express their contempt through lazy, inadequate responses to commands or blatant acts of disobedience), and owners — unable to prevent dogs from running out in traffic or assaulting mailmen — lose dogs. Dog love, in this view, requires tough love.

At the other end of the spectrum is the philosophy you'll find in books with titles like *Training with Love, Praise, and Reward* and *Dog Training: The Gentle Modern Method.* This is the dog-as-partner view, or dog-as-slightly-retarded-child, a view of the dog as an innately good, sweet-tempered, loving creature who needs guidance and positive reinforcement rather than punishment and force. Here the coercive, physical methods advocated by trainers like Koehler

— lots of firm corrections with a choke collar, for instance — are seen as punitive and unnecessary, and the canine experience of love (or what we can know of it) is believed to be largely compatible with our own: you can, in fact, adore your dog into submission.

That's certainly an appealing view, but it's not very widely shared. The third and largest camp comprises more mainstream trainers like the Monks of New Skete, Brian Kilcommons, and Matthew Margolis, who exist somewhere in the middle of the spectrum, acknowledging the dog's potential for both aggression and goodness, and doling out equal shares of firm physical corrections and enthusiastic words of praise. There in the middle, the human style of loving is recognized as both understandable and potentially dangerous, an asset and a liability: sure we love dogs, and well we should, but loving a dog isn't the same as controlling one, and what we perceive as sweet, nurturing behavior may be perceived by dogs quite differently. My trainer, Kathy de Natale, belongs to this camp, and I remember the look of barely disguised horror that crossed her face when I told her that Lucille, then four months old, was sleeping on my bed every night. Among many trainers, the idea of letting a dog, particularly a young pup, sleep with you is widely considered taboo because it gives him the message that he's an equal. So can allowing the dog to jump on you, or to growl at you if you approach his food bowl. The logic: being cuddly and permissive may feel good to *you,* but such behavior isn't necessarily good for your relationship. It may undermine your power and help the dog figure out whether he can successfully challenge your authority.

I also remember standing in an early puppy class, reading directives on a handout Kathy passed around: "Teach the puppy that nothing is free! Make him sit before you give him a treat, or before you pet him." The reasoning here was clear — if you make the dog earn things like biscuits and pats, you reinforce the idea that you are in charge, not him — but I recall a feeling of collision inside, intellect bumping up against emotion. Make her sit every single time I want to pet her? No free lunch *ever?* I stood there thinking about all those unsolicited treats and strokes, and I gulped, imagining the dog's respect diminishing with each successive Milk-Bone. One of the reasons controlling a dog is so complicated, one of the reasons it can make you feel like you're standing on such shaky ground, is that it poses so many internal challenges. You have to look at your dog (and into your own heart) and figure out where you stand on the power-and-respect continuum, which camp you belong to: What *is* this creature? A crafty, uppity beast or a toddler in a fur coat? You have to balance your need for a relationship that feels affectionate and giving with your dog's need for hierarchy and order. You have to figure out what it means not just to express love to a dog but to generate love in return.

And sometimes — no small feat — you have to be the heavy.

Observe Leslie, thirty-eight, owner of a two-year-old Wheaten terrier named Wilson, who has just absconded with a jogger's glove. Wilson is boxy and tan and soft-coated, a teddy bear of a dog who looks fluffy and sweet and perfectly innocent, but when he sees the jogger, he trots alongside him and just snatches the glove out of his hand. Leslie is mortified. She tries calling Wilson: no response. She tries yelling: "Drop it! Wilson, *drop it!*" Nada. She tries chasing him, and of course he thinks this is a hilarious game: she gets within grabbing distance of the glove, and he dashes off, a look of devilish glee in his eye. Finally an onlooker grabs Wilson by the collar and wrests the glove out of his mouth. The jogger, glove returned, gives Leslie an annoyed look and proceeds on his way.

Leslie is freaked out by this, even two weeks later. "It is *so* embarrassing," she says. The episode makes her feel humiliated and out of control, and she worries that Wilson's behavior reflects badly on her, that when he's unresponsive and ill mannered, he becomes not just a dog but something much more revealing: a four-legged embodiment of her fundamental inability to set limits. "When people stop you on the street and say, 'Oh, he's such a nice dog,' " she says, "it's like they're saying, 'Oh, and you must be such a nice person, too.' And when the dog is bad — well, you can't help but think that people are thinking you're a bad owner."

When Leslie first got Wilson, she was as ignorant about dogs as I was, and her definition of training consisted of two eensy words: "house training." She thought: You teach the dog not to go to the bathroom in the house, and that's that — trained dog. Yeah, right. True to his breed, Wilson is an exuberant, energetic dog — Leslie loves this about him, but it also causes her problems.

He leaps up on people, anyone who shows up at her door. When she takes him out for walks, he drags her down the street, and people are forever shouting out, "Hey, are you walking that dog, or is he walking you? Ha, ha, ha."

And yet Leslie cannot bring herself to control the dog — at least not the way most mainstream trainers tell you to. When Wilson was about six months old, Leslie took him to a trainer who taught her what's variously known as the leash correction or the "pop," pretty standard training fare. Take one choke collar, issue a command, and if the dog doesn't obey, give the leash a good snap, hard enough to communicate to him that he's done something wrong. This method never bothered me — dogs are physical creatures who understand and respond to comfort and discomfort; they're also a good deal tougher than some of us give them credit for — but Leslie couldn't stand the idea. "It made me so uncomfortable," she says. "The trainer was so rigid and controlling. For me, the idea of relating to a dog by snapping his neck was like an anathema. I'd never relate to anybody that way." She lasted through one session, never went back. Leslie tells me about this, talks about how uncomfortable all that authoritarian neck snapping made her, and then she lowers her voice and confesses: "What I really want to do is *negotiate* with my dog."

Beneath that simple statement — "I want to negotiate with my dog" — are all those vexing questions about the nature of animals, about relationships, and about giving and getting love. Leslie is an art therapist who works with children in a local hospital, and although she knows Wilson is a dog — she can see that, he reminds her of it daily — a part of her genuinely wants to treat him like one of her young clients, like a young, vulnerable child. Leslie is also a great believer in equality in partnerships, in forming alliances: when she interacts with Wilson, she envisions, she says, "a kind of give and take," a dynamic that includes mutual respect and a willingness to compromise. And as she is discovering, she is not by nature an authoritative or domineering personality. "There's such a huge emotional overlay to setting limits, to being controlling," she says. "It makes me really uncomfortable to have to be that way." Authority, in a nutshell, conflicts with her relational style, with her concept of a loving bond, with her sense of self as a loving person.

Leslie, of course, is hardly the first dog owner in America to experience this conflict. "Thirty percent of my clients cannot assume a leadership role. Their definition of love just won't allow for it." Myrna Milani, D.V.M., a veterinary consultant and author based in New Hampshire, utters those two sentences with so much frustration, she practically spits. She is not alone. When Brian Kilcommons, nationally known dog trainer and author of several training books, starts talking about the human inability to set limits with dogs, you can practically hear his blood pressure rise: "People tolerate behavior in dogs they would *never* tolerate in a human being," he says. "I mean, you don't let your boyfriend come up and punch people in the crotch to say hi, or maul them because he doesn't like them. But that gauge is not applied to pets. The dog leaps up on someone, and the owner says, 'Oh, he's just being friendly,' or the dog attacks the mailman, and the owner says, 'Oh, the mailman must have done something to make him angry,' or the dog doesn't respond to a command — an animal whose hearing is far more acute than ours is — and the owner says, 'Oh, well, maybe he didn't hear me.'" Jodi Anderson, a trainer in New York, describes men who head up multimillion-dollar corporations, men who wield vast amounts of power, who are charged daily with the hiring and firing and control of other human beings, and yet who cannot bring themselves to reprimand the dog, cannot so much as put the dog in a simple sit-stay and ask him to wait for his dinner.

This all sounds rather amusing in a pathetic sort of way — you picture the big powerful CEO, a tiger at work and a marshmallow at home, held hostage in his own kitchen by a Yorkshire terrier — but lack of leadership can have fearsome consequences. A dog's mental health, after all, depends to a large degree on leadership: dogs get enormously distressed when they think no one is in charge. Accordingly, it's not only nonsensical to fail to establish rules and limits with a dog — to train him, to correct him when he's wrong — but cruel. Lack of leadership also ruins a lot of canine lives — and quite literally. In the absence of control, dogs get loose, they don't come back when they're called, they bolt into traffic, they die. (Staffers at Angell Memorial Hospital in Boston, one of the best-known veterinary hospitals in the nation, see this happen so regularly, they've coined a phrase for the phenomenon: dogs who get hit by cars are called victims of "heavy-metal disease.")

Those who aren't killed in city streets may

be killed at animal shelters. Four to six million animals are given up to shelters each year by their owners, and a healthy proportion of those — as many as 40 percent — are surrendered by disillusioned or frustrated people who didn't realize how complicated living with a dog can be. Numbers like these will break a trainer's heart. Often shelter dogs are young animals (about 25 percent of pets are destroyed by the time they reach two years of age) whose lives might have been spared had their owners put some time into basic training. A recent study from Purdue University found that people who hadn't taken their dogs to obedience classes were about three and a half times more likely to surrender the animal to a shelter than owners who had: training, not surprisingly, creates both a better-behaved dog and a closer bond between dog and human.

And yet the mindset and skills required to control a dog don't come easily to a lot of humans. You have to be tough. You have to deal with conflict in much more direct and physical ways than many of us are used to. Dogs themselves are enormously clear with one another; if you've ever watched a documentary about wolves and seen how a mother reprimands her cubs, you get an idea of the kinds of signals they respond to.

She cuffs the cubs across the snout, bares her teeth at them, nips them on the flank; she's clear and physical and harsh. In human form, "alpha" behavior — firm, stern, consistent — may not only feel unduly harsh or dictatorial; it may also evoke any number of unpleasant associations on the part of the owner, reminding us of the authoritarian parent or schoolteacher, of ancient feelings of being unloved or treated unfairly; as it did for Leslie, controlling behavior may run counter to our most basic ideals of what nurturance is supposed to look and feel like, particularly when it's directed toward an animal you adore.

My friend Wendy describes the first weeks with her new puppy, an Australian shepherd named Alaska, as a frenzy of proactive discipline. Alaska, now almost three, is a beautiful example of her breed — a forty-five-pound tricolored shepherd with a silky brown coat, white and tan markings, and green eyes — and she was as cute as they come as a puppy, a floppy brown puffball. When Wendy looked at her, all her instincts said: Melt! Cuddle! Pet! All the experts said something else. She remembers, "All the books tell you to *set limits* and *establish the rules* and *be the boss,* so you're running around saying 'No!' 'No!' 'No!' from the very beginning." Wendy compares this to her experience as a new mother some twenty-five years earlier, when her baby spent most of the time asleep, allowing her to indulge in the cuddling, bonding side of new motherhood. "Getting a puppy is more like acquiring a two-year-old," she says. "It's very disconcerting." Indeed, control can fly in the face of all those warm maternal instincts; it can make you feel withholding and mean, inspire guilt around even the most seemingly straightforward decisions.

Wendy and I talk about this, and a single image floats to mind: the crate. The image of tiny Lucille at the age of three months, barricaded in her wire crate, dark puppy eyes blinking out at me. The horror! The agony! And (on my part) the abject stupidity. Like most contemporary dog owners, I'd been encouraged by virtually every dog-training book I read to crate Lucille as a puppy, and I understood the rationale — because dogs don't like to soil the area where they sleep, crates are masterful tools for housebreaking; crates also prevent the puppy from tearing up your house, and they give the puppy the safe, enclosed sensation of being in a den. So I trooped out to buy one, dutifully stationed it in my bedroom, and proceeded to freak out. Reason is no match for the sound of a crying pup. I'd put Lucille in the crate before I went to sleep, and she'd promptly start to wimper, and I'd lie there in bed and want to die. *She's miserable! She's lonely! I'm causing her pain!* Invariably, Lucille would settle down after a few minutes and fall asleep, but I found the act of distressing her for even a moment excruciating. If she woke up in the middle of the night needing to pee, I'd take her outside and then let her spend the rest of the night with me on the bed. On the third night of this, I woke to the sound of her crying, stumbled out with her, then got back into bed and looked at the clock: 11:30. We'd been asleep for less than an hour. Dogs are very smart. The crate was history.

I came to regret caving in so quickly — all the dog owners I knew at the time swore by crates, and all their dogs still seem to love them, creeping into them for naps of their own accord — but the experience was telling, a testament to that dark battle between intellect and feeling. I'd put her in the crate, and some part of me would feel convinced, against all reason, that the act felt cruel and punitive to her; some part of me would worry that she'd love me a little less when she got out.

But what can we really know about love? Would Lucille really have loved me any less if I'd made her stay in that crate at night for three months instead of three days, or if I'd forbidden her to sleep on the bed altogether? Would she really love me any more if I were more of a hard-ass, more controlling and forceful?

When Lucille was about fourteen months old, a Boston-area behaviorist named Jay Livingston looked me straight in the eye and told me: "Your love and respect for Lucille far outstrip her love and respect for you."

Apparently this statement was based on about three minutes of observation. I'd taken Lucille with me on an interview with Livingston, and toward the end of a general conversation about people and dogs, he told me he could pretty much scope out a human-dog relationship that quickly: within minutes. He also said his initial assessment is almost always right: in formal consultations with clients, he spends the rest of the time merely confirming his perceptions. Intrigued, I asked him: "So how would you assess my relationship with Lucille? In the first three minutes of watching us interact, what did you see?"

Livingston hails from the uppity-dog

school of training, and he spoke at great length during our talk about how spineless modern dog owners can be, about their refusal to muster up the emotional strength, control, and pushiness they need to effectively deal with dogs. Given his bent, I didn't expect him to say he saw Lucille and me as a paragon of perfect training, but I also sat there as he talked feeling rather proud of my dog, and of the work I'd done with her.

I scratched notes on a pad, and I thought about the dozens of times I'd seen owners in the park screaming "No!" at their dogs in vain, or watching helplessly while their muddy-pawed pets leaped onto strangers or menaced other dogs. I considered some of the more idiotic things I'd seen and heard about in the world of dogs (a woman with a highly aggressive pit bull who called a trainer seeking advice not about curbing the dog's behavior but about opening a bed-and-breakfast in her house; a woman who arrived at the dog park one day boasting that her four-month-old puppy had followed her, unleashed, *all the way through Harvard Square!*). And after a while I even began to feel a teensy bit smug. In the world of dominance and control, I'd hardly be described as dictatorial, but I'd worked hard with Lucille at obedience training in the year since I'd gotten her, and despite our periodic battles over the recall, her behavior seemed to reflect that. While I talked with Livingston, Lucille spent most of the time lying near my chair on the floor, sleeping, and I'd look down periodically and think: What a sweet pea she is; my good dog.

So I expected Livingston to assess our relationship in rather benign terms: "You seem to be doing well together," or "It's hard to say, given that I haven't really seen you working together." But Livingston sat back in his chair. "Well," he said, "it's obvious that you've done some basic, AKC-style training with her. It's obvious that she's attached to you." And then he paused and said it: "And it's obvious that your love and respect for her far outstrip her love and respect for you."

What astonished me about this summation was its power, the vulnerability it unleashed in me. I listened calmly to Livingston's explanation. Lucille, he said, is a dog who comes from serious working stock, and in his view she "wasn't doing any work at all." When we came into his office, he watched her sniff around the perimeter of the room, jumping up on her hind legs at one point to investigate some toys on a high shelf. "She's fourteen months old," he said. "You may think, 'Oh, she's just being a puppy,' but I see an adolescent dog who should know better. That's obnoxious behavior to me." I nodded soberly as he talked about canine ideals. Dogs function best, he said, "when they're well loved and highly controlled," and he saw something missing in part two of that equation: Lucille was paying more attention to her surroundings than she was to me; instead of wandering about on her own, she should have been far more focused on me, ready to drop into a down-stay at a moment's notice. I asked a few vague questions ("What kind of 'work' should she be doing?" I wondered. His answer: "Pleasing you."). And then I gathered up my notebook and my dog, went home, and burst into tears.

I eyed Lucille warily for the rest of the day. She doesn't love me as much as I love her? Doesn't *respect* me? This dog, whom I so adore, thinks I'm a wuss? My oldest, most familiar fear welled up: You're unworthy of love — even your dog knows this. It took me several days (and several conversations with my own trainer) to depersonalize his assessment, to accept that his ideal of control is simply different from mine, to acknowledge that ultimately all dog owners

130

have to find the right balance for themselves. But the experience was striking all the same, a reminder of the extent to which the matter of control can get tangled up with ego and insecurity, with our most central worries about love.

Livingston's analysis might have had more to do with human aggression than canine love but, inadvertently perhaps, it may have inspired me. Around the time of that interview, I decided I couldn't tolerate standing there at Fresh Pond screaming into the bushes for my dog — the episodes scared me too much, made me too angry, generated too much discomfort on the love-and-respect front — and so I embarked on a long remedial program, the human-dog equivalent of "working on our relationship." Every day for a month I'd gather up a thirty-foot leash and a pocketful of freeze-dried liver, and I'd take Lucille to a soccer field just off the Fresh Pond walking trail. I'd place her in a sit-stay, walk twenty feet away, turn and face her. Then I'd summon up my most upbeat energetic voice: "Lucille, come!"

She'd look away.

I'd say it once more, with feeling — "Lucille, *come!*" — and if she failed to respond again, I'd give the leash a good snap and reel her in to me like a fish. She'd come plodding toward me at the other end looking vaguely irritated, perhaps distracted, but I chose to ignore this. When she finally got to me, I'd practically fall all over her with enthusiasm: *Good dog! What a good come! Excellent!* Then I'd hand her a cube of liver and begin again.

We repeated this exercise ten times, twenty times, thirty times a day. Sometimes I felt like a Marine sergeant, and sometimes I felt like an Olympic coach, and sometimes I felt supremely bored, but I kept at it. And it worked. By the end of the month, Lucille would drop dead carcasses in the woods to come to me. She'd ignore tempting smells and other dogs; she'd tear out of the brush and plant herself at my feet; my heart would swell with pride and relief.

So did I become Master of the Universe in her eyes? Did she develop new respect for me? Or was it really all about liver?

Who can say for sure? Lucille has developed a fairly consistent recall rate — she'll come when I call about 90 percent of the time — and I, in turn, have developed a more comfortable relationship with control. Not an ideal relationship, mind you. Some part of my ego may always be drawn to the Master of the Universe role, and I can still experience a little stab of jealousy when I see someone whose dog is perfectly trained and unwavering in focus, but I've come to terms with the idea that I have neither the will nor the need to be that highly controlling. This means compromising: I let the dog off-leash in the woods, keep her on-leash in the city; I worry periodically about whether she "respects" me sufficiently, but I no longer worry excessively. (And the lingering sting from Jay Livingston notwithstanding, I don't really believe that Lucille is lying awake nights plotting ways to overthrow the regime.) In other words, I've developed a level of control I can live with, one in which both my ego and her safety are protected. But I am still struck — often, and powerfully — by how much I don't know about what goes on between us, about how she really perceives me or what really motivates her responses. I have been able to shape her behavior to some extent, to learn how to predict it, to be more comfortable with it. But controlled or not, bounding toward me or bounding away, she is still the dog: mysterious and, at heart, unknowable.

5

INSCRUTABLE DOG

The dog appears to be practicing telepathy. We are at a friend's house for the evening, and she keeps creeping into the kitchen where we're drinking coffee, planting herself about four feet from my chair, and staring at me, her gaze unwavering and intent.

I look at her. "What is it, Lucille?"

She stares and stares, ears pricked, body taut with concentration.

I pose the obvious question: "Do you need to go out?"

She doesn't move, which suggests that this is not the issue (if it were, she'd jump up from her sitting position at this point and do a little prance, code for *yes-yes, that's it*), but I leash her up and take her out anyway. She pees but expresses no great urgency about it, and ten minutes later she's back at it: comes into the kitchen, sits, stares.

This continues off and on for about ninety minutes. I reach over to pet her periodically, tell her to lie down. She complies periodically but invariably gets back up and stares some more, clearly struggling to request *something* of me. My inquiries lead nowhere. "Do you want a toy?" She doesn't blink. "Do you want to go home?" Doesn't budge. She looks like she's telling me she needs to *be* somewhere — an appointment, an important date — and I am struck, as I so often am, by the communicative gulf between us, how massive and unbridgeable it can feel.

J. R. Ackerley describes this in *My Dog Tulip*, his tender 1965 memoir of his beloved Alsatian bitch: "What strained and anxious lives dogs must lead," he writes, "so emotionally involved in the world of men, whose affections they strive endlessly to secure, whose authority they are expected unquestioningly to obey, and whose mind they can never do more than imperfectly reach and comprehend." At the moment Lucille is reaching my mind imperfectly indeed — I have no idea what she wants or needs or feels — and I think about Ackerley's words as I look at her, about how stressful it must be for the dog not to be able to speak English.

It's stressful for humans, too. Living with a dog — trying to understand a dog, to read his or her behavior and emotional state — is such a complex blend of reality and imagination, such a daily mix of hard truths and wild stabs in the dark. Lucille and I have moments of genuine sympathy, small daily exchanges that make me feel we bridge that gap and know one another, to the extent that this is possible between members of different species: a look here; a touch there; words uttered by me, understood and reacted to by her. But for each instance of clarity, there is a moment that baffles me utterly, like the episode in my friend's kitchen. Does she want something? What is it? Is she unhappy? What is she *feeling?* The gaze betrays nothing; her internal state is alien to me. And so I do what a lot of dog owners do: I stumble through this unknown territory, sometimes blindly and sometimes not, sometimes accepting the gulf between us and sometimes resisting acceptance, and often — often and inevitably — falling back on my own entirely human interpretations.

In other words, I project. I project and I anthropomorphize and I make stuff up. I view her inner life through the filter of my own emotions and experiences, and the tendency to do this can make me crazy, for I can read anything into Lucille's eyes. *Anything.* I can imagine that she's mad at me, whether or not she is. I can imagine that she's lonely or depressed, that she's worried or chagrined or wistful, that I'm getting on her nerves. Not long ago I rattled off a list of emotions I've attributed to Lucille to a friend, someone who doesn't have a dog and has no emotional investment in dogs. She heard the list (melancholy, regret, anticipatory dread, spite), and then she smiled. "You know," she said, rather gently, "her feelings are probably really *simple*. Like, Bed: comfortable. Or, Need to pee: uncomfortable."

My friend is probably right — at the very least I suspect that Lucille's experience of emotion is different from mine, that her inner life is immediate and sensate and less associative than mine, cleaner somehow, certainly less abstract. But accepting that doesn't dampen my curiosity (I'm *dying* to know what goes on in there), and it certainly doesn't put the brakes on my imagination. The fact is, the dog is a blank screen. Lucille appears to feel very powerfully at times, and she can seem nearly human in expressing her moods, her eyes communicating excitement or glee or desire as clearly as my own do. But the true nature of her emotions — how she feels — is a mystery to me. Lucille doesn't have the words to describe them, and so I'm left with nothing but shadowy guesses, vague speculation about the canine mind, ideas that are shaped (sometimes consciously, often not) by my own sense of what it means to have feelings. This is the nature of the beast: dog as Rorschach. Lucille looks at me with those big, dark eyes, that empty canvas of a face, and I can read into her expressions any number of feelings — sadness, disappointment, empathy, love. And because she can't talk back, because she can't challenge my interpretations of her emotional state, I am left to grapple with what I see on my own.

Consider, for example, the canine gaze, Version II:

I am at home with a group of friends, eating Chinese food in the kitchen, and once again Lucille has stationed herself next to my chair. This time the gaze is plaintive and earnest, and this time I know precisely what it means. She is imploring me, beseeching. If there were a thought balloon over her head, it would read: "Give me a bite of that crispy beef. Give me some Szechuan chicken. You're eating; won't you feed me, too?"

I feel a little stir of anxiety inside, tinged with guilt. I hate depriving the dog of something she so clearly wants, hate that beseeching gaze, but I refuse to feed her from the table: that's one line I refuse to cross. Most nights, when we're at home alone, I placate her with a special recipe: I take two hollow bones, stuff them with dog food, and give them to her when I sit down to eat, a solution that serves both of us (she gets to lick the food out of the bones, a pleasurable chore that takes her as long to complete as it does for me to finish my own supper; I get to eat without being stared at). But tonight is a variation from the routine: new foods, new smells, new people, and (I suspect) a sense of new possibilities for the dog. So she stares. She begs and beseeches, and finally I surrender, get up from my chair, and fetch her a rawhide stick from a drawer. She looks thrilled and relieved at this — *at last! success!* — and she darts out of the room as though carrying something illicit, the stick between her teeth. But ten minutes later she's back at my chair. Sits there, stares and stares. And I am back in conflict. I am back on the slippery slope of projection, the place where behavior and persistence (hers) meet emotion and interpretation (mine).

There is a quality of such helpless innocence to the dog: their lives are so shaped and constricted by the decisions we make for them, and I don't know a single owner who doesn't struggle under the weight of that responsibility, at least to some degree. The awareness of this strikes me as I sit there at the table, my begging dog beside me. However we feel about our status as pack leaders, we are in charge of these creatures and every feature of their existence is at our whim: we determine where they sleep; how often they go out and for how long; how much time they spend alone; whether they play with, mate with, or even see other dogs; whether they'll get scraps of food from the table or go without. These are broad areas of life — sexuality, social interaction, food — and they are areas about which both humans and dogs appear to have very strong feelings. This is why anthropomorphism (attributing human emotions to dogs) and projection (attributing your own emotions) can feel like such inevitable impulses, why they're so difficult to resist: when you dictate the terms of another being's life — a voiceless animal, whom you want to tend to properly — it can be supremely hard to keep your own emotions under wraps, to keep them from coloring your view of the dog's experience.

At the moment Lucille wants something to eat: this is a simple fact, and it may be, in her mind, a simple experience, devoid of emotional associations. *Food. Interesting new smells. I want.* A poem by Karen Shepard, written in the voice of her dog, Birch, captures this possibility in four perfect lines:

Are you gonna eat that?
Are you gonna eat that?
Are you gonna eat that?

I'll eat that.

But eating — feeding and being fed — is not nearly so straightforward to me. In my mind, food is loaded with meaning, and the giving and withholding of food are profoundly emotional acts. If I give, I am being loving and generous and nurturing; if I don't, I am being cold and mean and the dog will think I don't love her.

These are not logical thoughts but the feelings behind them are deeply rooted, shaped by years of my own experience with deprivation and nurturance, hunger and satiation. Like a lot of women, I've channeled oceans of longing into a preoccupation with food over the years, danced elaborate dances with craving and reward, used food to fill some nameless internal emptiness, and all of this can come surging up when I see those imploring eyes; my own associations are nearly impossible to ignore. On this particular night I do my best to avoid eye contact. I stare at my plate, talk to my friends; I give the dog nothing else. Finally Lucille surrenders the vigil, retreats to the living room and curls up on the sofa. I watch her creep out of the room, and I remind myself that, to her, deprivation is probably no big deal: she may be profoundly interested in my plate of crispy beef, but chances are she's not associating it with inchoate hungers or existential yearning. And then I return to my dinner, trying to ignore the lingering, not terribly rational worry: Is she out there in the other room feeling dejected and deprived? Have I hurt her feelings? I've won the battle, but just barely.

Dog owners wage battles like this all the time; they're a central feature of living with a dog, and food — so tied to feelings of longing for humans, so wrapped up with our concepts of nurturance and love — probably sends us into more projective tizzies than any other issue. I've talked to owners who toast their dogs bagels for breakfast, spread them with cream cheese and raspberry jam, set them down on the floor on special plates; I know owners who fix their dogs individual bowls of ice cream every night, a different flavor for each day of the week, sprinklings of jimmies included; I know owners (I'm one of them from time to time) who spend more time preparing the dog's dinner than their own. Some of this is genuinely loving behavior on the humans' part — there's an intimacy to the act of feeding the dog, at least for us — but if we didn't have such complicated feelings about food ourselves, chances are we wouldn't go to so much trouble.

Dogs, of course, fuel the projective fires; they play us like fiddles. As behaviorists will tell you, dogs are often far better at training us than we are at training them, both clearer and more consistent than humans in communicating their wants. They are also very quick studies when it comes to food: they learn which behaviors are rewarded and which aren't, what they're likely to get away with, what will or won't happen if they refuse to eat what we put in front of them. I know a woman whose Doberman pinscher has basically "taught" her that he won't eat his dinner unless it's mixed with two tablespoons of extra-virgin olive oil. No other oil will do — try to fake him out with low-grade peanut oil, and he'll sniff the bowl and walk away. Sarah, his owner, knows this is a little silly — she picks up that twenty-eight-dollar bottle of olive oil every time she feeds the dog and she thinks, Ridiculous — but the olive oil gets the dog to eat, and the owner can't stand it when the dog goes hungry.

And so the ritual is repeated, day after day.

Brian Kilcommons and Sarah Wilson write about a family who projected themselves right out of their own home during feeding time. Somehow — they never quite figured out how this happened — their dog would eat only if the entire family trooped out to the hallway outside their apartment and rang the doorbell. If they failed to enact this ritual, the dog would ignore his food and go hungry, causing the loving, nurturing humans great distress. So they'd file out the door, ring the doorbell, and stand there while the dog ate. A little nutty? Perhaps, but I'm hardly the best judge. I've been known to sprinkle Lucille's food with liver powder and Parmesan cheese when she won't eat, to mix it with chopped carrots and expensive Italian pasta, to get down on my knees and *hand-feed* her. What she makes of all that effort is beyond me, but it gets the dog to eat, which (in my mind, anyway) is the point.

These are the kinds of stories that make vets and behaviorists gnash their teeth. C'mon, they say. The dog is the dog. The dog doesn't need us to jump through such elaborate gastronomic hoops. The dog doesn't need fancy food or varied food or human food or health food, and he certainly doesn't need food that comes in pretty little shapes and colors, the stuff that marketers dream up in order to appeal to the *owner's* definition of an appetizing, healthful diet. Dog biscuits shaped like mailmen and cats? You think the dog really knows the difference? The same could be said of some of the products we buy for dogs, such as beds designed like little futons and canine apparel. One rainy day last fall I saw a perfectly groomed white standard poodle standing outside in a field, clad in a pair of blue plastic rain pants. Think that was the poodle's idea?

It's not too hard to guess how a poodle feels about being swathed in Gore-Tex, but fathoming the canine experience on more complicated fronts can be very difficult, indeed. Leslie, owner of the Wheaten terrier named Wilson, struggled for nearly a year with the question of whether to neuter her dog, a concern that was wrapped up in her own feelings about sexuality, her own sense of what it meant to be a sexual being. When Wilson, not yet neutered, was about a year old, Leslie saw the movie *101 Dalmatians*. She watched the two protagonist dogs fall in love. She watched them marry and have a family and coo together over their puppies, and this image crystallized certain fantasies she harbored about Wilson's life. Leslie, of course, understood that the movie was a Disney production, warm and humanized. She understood that concepts like marriage and family unity are alien to dogs. But the movie tugged at her, and part of her wanted what it represented: sexual love for her dog, the opportunity for him to express that part of his being. Neutering seemed like such a betrayal by contrast, cruel and irrevocable, and the idea of tampering with his most basic drive, his sexuality, horrified her.

This fear tends to be more common among men (according to Alan Beck and Aaron Katcher, pioneers in the study of the human-animal bond, some men have such a hard time separating the concept of virility from the physical equipment that they say they'd rather have the dog killed than neutered), but Leslie agonized over the decision: she made several appointments, canceled them at the last minute. Finally Wilson started bolting out of local parks and into traffic, drawn by the scent of females in heat; worried about his safety, Leslie rescheduled the procedure. I spoke to her the day before the surgery. She felt it was the right thing to do by then, and she was determined to go through with it, but she never fully came to terms with the idea that for

Wilson, sexuality was probably more a physical, hormonal experience than a romantic one. Shortly before I left, she told me: "I've been thinking that maybe I should get a female dog. Then Wilson could get a vasectomy and I could get the female's tubes tied, and then it would be an equal thing. They could stay home while I go to work, and they could have sex whenever they wanted to."

Nice fantasy, but not necessarily a very canine one. When I suggested that the dog's experience of sexuality is probably not as tied in to intimacy as ours is, Leslie sighed and said, "Yeah, I guess you're right. I just have such a hard time seeing him as a dog, and not as a person."

Nearly half of all contemporary dog owners appear to struggle with that distinction: 48 percent of respondents to a University of Pennsylvania survey said they saw their dogs as "people" rather than as dogs, and I'd venture to say that even those of us who are very clear on the fact that dogs are dogs blur the lines between us and them on occasion.

For Milton, a forty-four-year-old psychiatrist, the projective downfall used to come in the first ten minutes of puppy kindergarten class, which was designated as playtime, an opportunity for the dogs to socialize. Milton was new to the world of dogs, having just acquired a boxer puppy named Zelda, and he'd never really seen dogs play before. They'd start hurling each other to the floor, baring their teeth, executing hip checks, and Milton would stand there in abject horror, not because he feared for Zelda's safety (mock fighting is perfectly normal canine behavior, as natural to a puppy like Zelda as chasing a ball; Milton knew this), but because the sight rocketed him back to boyhood, to the vulnerability he had felt in seventh-grade gym class as "the smallest, runtiest, least athletic kid in history." He says, "I'd stand there with my heart in my throat thinking, 'The other kids are beating Zelda up! The other kids are beating Zelda up!' All that old fear — *my* fear — came washing up." For her part, Zelda turned out to be one of the more dominant dogs in the class, perfectly capable of holding her own against seven other puppies, but Milton still shudders at the memory: "Playtime," he says, "*terrified* me."

Dog as small, fur version of the owner as child: very easy view to slip into, very hard to resist. I'm better at separating Lucille's experiences from my own than I used to be, but when she was a puppy, I had to give myself little lectures all the time, little mental slaps to tease out what I knew about her world from what I felt about my own. I'd see her at the dog park, standing hesitantly at the outskirts while a group of dogs played, and my heart would melt: the canine wallflower, tentative and self-conscious, blinking out at the other kids. Mental slap; private lecture on popularity, human versus canine: The dog is not feeling left out; dogs are not preoccupied with such human concerns as "fitting in"; the dog simply doesn't feel like playing. Or she'd fail to execute a command at obedience class, and I'd look at her, certain for the briefest moment that I saw embarrassment in her eyes, and then I'd have to give myself a lecture about human, as opposed to canine, definitions of failure and success: Lucille is not worried about getting an A here; the concept of performance anxiety is alien to the dog; this discomfort is entirely yours.

What's astonishing to me is how automatically such feelings can arise, how unconsciously they attach themselves to the dog. Betsy, a fifty-two-year-old New York editor, describes taking her Tibetan terrier, Lucy, then four years old, to Central Park, where she ran into a young woman with an adorable puppy. The two chatted. Betsy asked the woman how old her dog was. Nine months. The woman asked Betsy how old Lucy was. And Betsy, who's never once lied about her own age and is determined not to, heard herself say, without even thinking, "She's *two*." Then she stood there and wondered: What was *that* about? Projection of her own fear of aging onto the dog? Competitiveness on the who-has-the-cuter-pup front? A statement about her own self-image? Whatever its source, *something* came up — some vulnerability, some buried insecurity, some need to manipulate the truth — and Lucy served as a lightning rod for the feeling.

During our first year together, I used to take Lucille once a week to a day-care center for dogs. The center is housed in a ramshackle triple-decker in Somerville, Mass., a town over from where I live, and when you walk in, all you see is *dog*. Two cavernous rooms with dogs everywhere: dogs on old sofas and dogs on dog beds; dogs lying on the floor, dogs waiting by the door; big dogs, little dogs, medium-size dogs; dogs, dogs, dogs. The place terrified me, not because I found the dogs scary but because I assumed — instinctively, unconsciously —

that Lucille felt lost and invisible when I left her there, exactly the way I'd feel if deposited in a crowd of strangers. I'd walk in, holding my treasured dog by the leash, and I'd think: Oh, no! She's going to disappear into this sea of dogs! Some part of me — a part that wanted to stand out even as a kid, a part that always felt lost and uncomfortable in a crowd, a part that lacked social confidence — would come welling up, and I'd practically burst into tears, the idea of her vanishing into anonymity like that so upset me. Rational? Not necessarily. Nothing in Lucille's manner suggested she was afraid or lacking in confidence or "feeling invisible" in the midst of all those dogs, but I'd see her watch me from the window as I left, see that dark face peering out after me as I got into my car, and I'd want to die.

The great frustration, of course, is not being able to *ask* the dog: Lucille, do you *like* day care? Are you happy being left here or would you rather be at home? Day care turned out to be a classic example of the way projection can color the decisions we make for dogs, making the question of what's good for them enormously difficult to address with clarity. Despite my anxiety about abandoning Lucille to that sea of dogs, I continued to take her to day care for about six months. The ostensible reason was to free up a chunk of time for myself, but I've come to wonder if I wasn't also motivated by a projected belief: that she'd be happiest in the company of other dogs, that day care would be "fun" for her. When she was fourteen months old, I took her to a camp for owners and dogs for largely the same reason: it sounded like canine (and owner) bliss — a week of fun in the country, new friends and activities for both of us. In retrospect, both services feel a little dicier than I'd originally thought, based more on human wishes, and human definitions of camaraderie and entertainment, than on canine ones.

Certainly they both cater to human fantasies: at one Boston-area day-care center, employees call the dogs "kids" and joke that they have "pillow fights" at night; at another, dogs who are boarded overnight are said to be attending "doggie pajama parties." The anthropomorphic slant at camp could be found in the itinerary: along with such standard fare as obedience and agility training, it listed activities like "doggie square dancing," a tail-wagging contest, a weenie-retrieval competition. Are these the kinds of things dogs really like to do? Well, not my dog. Dogs are creatures of habit and routine; they tend to dislike change, especially as they get older. A lot of dogs also find it very stressful to be housed for long periods with groups of unfamiliar dogs, particularly if they're high-strung or shy. The day-care center I went to served upward of thirty dogs a day, and as calm and adaptable as Lucille is, I'm not at all sure she found the experience "fun"; when I'd pick her up at the end of the day, she'd pass out in the car within minutes, and her fatigue seemed different from the kind she exhibits when she's been out playing in more familiar settings, more psychic than physical. At camp the population hovered around ninety and the experience wore Lucille out, not because she had such a rollicking good time square dancing and retrieving weenies but because she seemed to find being around that many other dogs exhausting, wildly overstimulating.

I don't want to overstate the projective dangers here: day care is an excellent option for some dogs, particularly well-socialized, confident ones who are introduced to it at an early age, and it certainly relieves a great deal of strain for owners who are loath to leave their dogs alone for eight, ten, or twelve hours at a stretch. Camp may serve some populations very well, too (members

of the show-dog circuit, whose dogs are accustomed to being around large crowds,
seemed to be most at home; the staff also included some top-notch trainers, a boon to
owners who don't have access to good
training in their hometowns). But Lucille's
stress was instructive to me, a lesson in how
much confusion the gap between our views
and their needs can generate.

One projected feeling leads to another,
then another; they nestle inside each other
like so many Russian dolls. At heart, I suppose my impulse to haul Lucille off to day
care and camp also sprang from another
anxiety, a suspicion I've struggled with
since Lucille was a pup: that she finds her
life with me inadequate, that she wishes she
had a more exciting, adventuresome owner,
that she's . . . bored.

This is one of my greatest relational fears,
that the beings I love will lose interest in me,
and I suppose it's no surprise it crops up
with the dog, blank screen that she is. I see
her look at me from her dog bed, her expression vacant and unreadable, and it's often
the first place my mind goes: The dog is
bored; she's sick of me, I'm sure of it.

I can make myself insane with this. Not
long ago, on a cold, raw, afternoon with a

wind-driven rain, Lucille and I skipped our customary outdoor outing and instead spent a couple of hours with my friend Tom and his dog, an Australian shepherd named Cody, one of her favorite playmates. Easy afternoon, fun for dogs and humans: Tom and I got to sit in his warm, dry apartment and sip coffee; the dogs got to wrestle in the living room, which they did for ninety minutes straight. Lucille appears to adore both Tom and Cody — every time we go to his house, she practically does a little jig on the front steps — and I kept stealing looks at her throughout the afternoon, gratified to see how happy she seemed. This is not always clear with Lucille — some dogs are naturally wiggly and exuberant; they wag constantly, wear looks of contentment nearly all the time, but Lucille's expression tends to be so serious and sober, her manner so quiet and composed, that she can appear to be in a grave mood, crestfallen or disheartened, no matter how at ease she actually feels. At Tom's she radiated glee: she'd come up to one of us and wag her tail, steal a kiss; she'd hurl herself at Cody; her eyes appeared bright and fascinated, the expression of a kid at a circus. And then, two hours later, we went home, she lay down on her dog bed, and she looked . . . miserable. Expression blank, not a trace of that earlier joy. My heart sank.

This is what I mean about driving myself insane. In reality the dog was probably worn out, the blank look connoting nothing more complicated than physical fatigue. But I saw wistfulness: *She misses Cody.* I imagined the little mind clicking with comparisons: she's thinking, *Cody's house is more fun than my house.* I heard a tone of sad resignation: she's thinking, *Ugh, here we are again, hanging around this same old living room; what a bore.*

These are not canine thoughts and feelings. I know this. I stood there and gave myself another round of mental slaps, a lecture on the experience of boredom, canine versus human. Dogs get bored (or more accurately, restless) when they're left alone too long, when they're underexercised or understimulated, none of which applies to Lucille (in fact, the opposite is true: I exercise her a fanatical two to three hours a day, never allow myself to skip a walk or cut one short, will lead her through the woods in monsoons; in my effort to stave off boredom, I tend to wear the poor dog out). Dogs also appear to relish their predictable, territorial comforts: lazing about on the sofa night after night may sound excruciatingly dull to me, but it's probably heaven to her. In all likelihood dogs do not make comparative assessments about their lives either, do not lie around wishing they were elsewhere, fantasizing about better owners, dreaming of more varied settings. The very idea that Lucille harbors such thoughts is absurd; I understand this. But that's the projective stew: human anxiety, a voiceless dog, a pinch of paranoia.

In other words, my stuff. I am deeply afraid of being bored: it's one of my least favorite states, almost always laced in my experience with undercurrents of alienation and despair, invariably a signal that I'm depressed. I am equally afraid (if not convinced on some level) that I am a boring person, hollow and bland, and that anyone who gets too close to me is bound to figure this out. The projected anxieties follow: when I see a glazed look in Lucille's eyes, I fear she's experiencing boredom exactly the way I do, with all its ancillary emotion. And when I see an empty, vacant look, I see something just as scary reflected there: me, my worst self.

Why would anybody stay with me if they had a choice? This is an agonizing fear, but it lives deep within me, and it's stubborn as a virus, immune to logic. When that feeling hits — she's bored; oh no, the dog is bored — all reason goes out the window. My fear becomes her reality.

I am my dog, my dog is me. Boundaries can blur like this in a heartbeat. Susan Cohen, director of counseling at New York's Animal Medical Center, calls projection "diagnostically fascinating" for precisely that reason. "When someone offers what sounds like a human interpretation of a dog's behavior," she says, "it gives you something to explore. It might not tell you a lot about the dog, but it helps tell you what the person is thinking, what they're hoping, fearing, or feeling." I tend (at least in my more difficult moments) to err on the fear side of that equation — I look into Lucille's eyes and see the reflection of my darkest fears — but I think it's just as easy to err in the opposite direction, to see much loftier sentiments beaming back from the canine gaze, whether or not they actually exist.

About a year ago I came across a worried message on an Internet dog chat line, posted by a forty-one-year-old mother from Florida named Marsha. Several days earlier Marsha had been feeding both her dog, a four-year-old Airedale terrier, and her baby, a five-month-old girl, scraps of pizza crust from the table, when all of a sudden the dog growled at the child, then snapped several times at the air. Marsha wanted to know: What was that about? Should she worry about it? Do something? Missives flew in response. Aggression alert! Dog is seeing baby as competition for the food source! Yes, you should worry (Airedales can be big, scary dogs), and yes, you should do something: obedience work, long down-stays, an end to joint feedings. Marsha was ambivalent about the response. In a subsequent e-mail dialogue with me, she said the incident seemed like a "one-time thing," and she refused to believe the dog would do anything to hurt the baby.

"He loves her," she wrote. "I really think he knows she's his little sister." Valid belief or anthropomorphic hope? For the baby's sake, one hopes the former, but Marsha's sentiments speaks to the wish — widely shared, if often dangerously misplaced — to see the dog as a four-legged version of our best selves, a fur-covered human who brings to relationships the same degrees of love, loyalty, and commitment we do.

That wish is manifested in many ways; you see undercurrents of it all the time. I'm reminded of a button, seen on dog-owner lapels of late: *Dogs are people in fur coats.* I'm reminded of a woman who came up to me at a dog park, beaming because she'd told her six-month-old Doberman puppy to guard her bag while she ran across the street to get some water, and the dog had obeyed, had actually stood there — unleashed — and (as she saw it) guarded her bag. I'm reminded of the owner of a German shepherd dog whom I used to see in the woods when Lucille was a puppy: his dog would lunge at Lucille every time he saw her and pin her against a tree, his hackles raised and his teeth bared, and the man would stand there and chuckle. "He's just being friendly," he'd insist. "He's just trying to get her to play."

The idea that dogs embody our own most admired and coveted qualities is very seductive. Dogs become extensions of us so quickly — we want them to reflect well on us — and they are easy to idealize. And so the temptation to see them through the lens of our own hopes — to play with the facts, to spin stories that support our relational ideals — can be enormously difficult to resist. The snapping Airedale is the devoted family member who shares our values; the six-month-old Doberman is the loyal and communicative partner; the aggressive shepherd is every bit as friendly and outgoing as we are. The fantasy here, the projected wish, is old and abiding: underneath that fur coat, the dog is a person who loves us.

I understand the impulse to romanticize the dog; I struggle with it myself. At least once a day Lucille comes up to me if I'm sitting quietly in a chair, and pokes her nose toward my hand. This is the kind of moment that can make me feel in tune with her, that seems to speak to a level of connection between us that I cherish: she is asking me to give her my hand, and when I comply, she will sit down next to me and spend several minutes licking it. She looks like a little fawn at a salt lick when she does this — her ears go back, her manner is gentle and concentrated, and she sometimes places a paw on my wrist to steady my hand, a gesture that feels delicate and tender and full of affection to me. But is it? I'd like to think that this is about love, that Lucille is deliberately seeking me out for a kind of canine kissfest, but it's equally possible that she's indulging one of her other affinities: for the taste of hand lotion. So which interpretation is right? I choose to believe that the hand-licking is about both moisturizer *and* me (they're my hands, after all), but I'm also aware that an element of choice is at work

164

here, that this is my read, based at least to some extent on my own investment in the idea of loving communion with my dog.

In the end that may be all we're left with: choice. In *Dogs Never Lie About Love*, a book that posits an exceptionally romantic view of canine emotion, author Jeffrey Moussaieff Masson describes what happens when he is out walking and one of his three dogs strays too far from the others. "I will notice that the other two stop and wait for their companion to return," he writes. "They look at me as if to let me know that this is the right thing to do, and that I should wait, too. They do not want to continue until the pack is complete." Some observers would interpret this behavior as strictly instinctual: keeping the pack intact is, for dogs as for wolves, a matter of survival. Masson chooses to see only benevolence: "This act," he writes, "is surely indicative of compassion."

Maybe, maybe not. The fact is, Masson will never really know if his dogs experience compassion the same way humans do, and I'll never really know if Lucille is motivated to seek out my hand by lotion or by love: they're dogs; they simply can't tell us. And at heart, I'm not so sure I'd want them to.

When Lucille was about a year old, I

started hearing rumblings about a practice called "interspecies telepathic communication," which is a fancy name for psychic communication with animals. I read a *New York* magazine article about a seminar on the practice, which attracted a standing-room-only crowd at a pet-care expo in New York; I saw a flyer tacked up at a pet shop, advertising a workshop on the subject; my dog trainer reported that more and more of her acquaintances were consulting "animal communicators" (the preferred term among practitioners) about their dogs and horses. The trade has attained the status of a minor movement in the last few years, and its practitioners answer human questions about animal emotion in unequivocal terms: Yes, they believe, animals most definitely experience emotions like ours, such as love and compassion. And yes, it is possible to reach across the boundaries of species and communicate with them quite directly.

Three hundred years ago this sort of thinking could have gotten you burned at the stake. The idea that animals even had feelings was seen as pure heresy at that time, and the Church (which controlled the bulk of research and scholarship) had a vested interest in keeping it that way. If animals were conscious, sentient beings — beings with souls — the Church would have been presented with a host of ethical problems: How to kill animals for food in good conscience if they have feelings and emotions? How to force them into servitude, deny them free will? How to make room for all those animal souls in heaven? It was far more palatable to see animals as unfeeling biological machines. Extending the ideas of René Descartes, the seventeenth-century philosopher credited as the first to propose that animals lacked consciousness, Nicolas de Malebranche summed up the prevailing wisdom, writing that animals "eat without pleasure, cry without pain, act without knowing it; they desire nothing, fear nothing, know nothing."

That philosophy may seem ludicrous, even cruel, today, but it persisted for generations, expressed in more modern form by the behaviorist school of the 1950s, which supported an equally mechanistic view: that animals were controlled not by "emotions" but by instinct, and by such empirically observable phenomena as neurons, muscles, and hormones; at least in the world of science, inquiries into an animal's inner life were roundly discouraged.

Today's scientists aren't exactly embracing animal emotion as a field of study (for the fairly pragmatic reason that animals can't tell us directly how they feel), but the general public certainly is. Far more personally invested in animals than we used to be, far more interested in relating to them as kindred spirits than in exploiting them, the contemporary pet owner would find a thinker like Descartes as heretical as he would have found us. We are deeply committed to the notion that animals feel, and we are starved for information about how they do it. In the past five years dog owners have made best-selling authors out of such emotionally oriented thinkers as Elizabeth Marshall Thomas (*The Hidden Life of Dogs*, published in 1993), Stanley Coren (*The Intelligence of Dogs*, in 1994), and Jeffrey Moussaieff Masson (*When Elephants Weep* and *Dogs Never Lie About Love*, in 1995 and 1997). And some of us, nudging the cultural pendulum one notch farther, are consulting psychics.

Intrigued, I called three communicators, less curious about what they might actually "hear" from Lucille than about the underlying conviction, the belief that they could hear anything at all. All three operated similarly. Prices ranged from $45 to $65 per session, and because the work, in the words of one, "is all conducted telepathically," I didn't need to bring Lucille in for actual visits. I merely called up, provided brief descriptions of Lucille and lists of questions, and they communicated with the dog, as it were, over the phone.

What struck me about these exchanges was their chatty matter-of-factness. I'd ask a general question ("Is Lucille happy?"), there'd be a pause of several seconds (time in which the communicator communed), and then an answer would come back, clear as handwriting on a wall. "Oh, yes," one said. "Her joy at being with you is, like, boundless. She doesn't even understand why you would question it." The same phenomenon applied to more specific questions. "Who are Lucille's favorite dog friends?" (Answer: "I'm seeing an apricot-colored spaniel.") "Is her diet okay?" (Answer: "She would like more vegetables.") The tone was unquestioning and absolute, as was the belief behind it: Ask and you shall hear; it's that simple.

This didn't feel like a scam to me. None of the psychics seemed crazy or deluded; there was an earnestness to their manner, a sense that each one truly believed she'd established some sort of contact with my dog. None of them made me feel particularly enlightened either, but for the most part, I found their interpretations pleasantly diverting, as I might find a reading of tea leaves. Boundless joy? That's perfectly pleasant to hear. Spaniels and spinach? Well, who really knows? I listened as they talked. I nodded. I scratched questions on a pad (*I wonder: Do Portuguese water dogs speak Portuguese?*). I didn't stock up on produce.

But not all the interpretations left me quite so amused. Example: Lucille has two phobias, which I asked all three communicators to explain. The first fear concerns highway driving, which she has loathed, for reasons that are not clear to me, since she was about six months old: she sits in the backseat and cowers, literally ducking every time we're passed by a truck. The second concerns houseflies, a fear that dates back to a week we spent in a rented farmhouse in Vermont when she was about a year old. The house was old and drafty, full of moths that would flit around the lamps at night, and my then-boyfriend, Michael, would periodically chase the moths around with a rolled-up magazine, smacking them against the wall in irritation. Lucille appeared to find this horrifying — you'd hear a smack, then turn around and see her slinking off to her dog bed. The housefly phobia materialized when we got back to the city: she'd see a fly buzzing around inside; a look of alarm would cross her face; she'd get up and creep away.

Two of the communicators had vague but not unreasonable explanations for these fears — the speed of highway driving may have upset Lucille; the houseflies may have reminded her of a bee that stung her that same summer; that sort of thing. The third, a woman named Marcia from Sherborn, Massachusetts, spun a more elaborate tale. "Oh, it's an awful story," she told me. In a former life, she said, Lucille had been owned by a mean "dark-haired, brown-skinned couple" who badly mistreated her. Lucille apparently showed Marcia a "picture" from this former life, a mental image of driving down the highway with the couple who, in a final fit of abuse, threw her out of the car and left her for dead. Marcia sent me a transcript of her "conversation" with Lucille: "While I lay dying," it read, "the flies were all over me, in my eyes and eating my flesh. They became so big. They were so destructive. They ate my body. I went through so much pain." Hence the dual phobia: highways and houseflies.

I read that transcript, thanked God I

didn't buy a word of it, and shook my head. Canine fears or human ones?

What troubles me about anthropomorphism, about projection that's sloppy and unexamined, is the idea that the elusive truths of a dog's experience are a mere phone call away, that we *can* know so easily and surely. One of the psychics encouraged me, toward the end of a consultation, to learn to speak with Lucille myself, to practice conjuring up mental images for her to receive and to make myself open to whatever pictures or sensations might be lurking in her consciousness. "Lucille says she would like to be able to communicate with you," she said. "She would be very willing to assist you. She's actually very enthusiastic about it."

I hung up the phone and looked down at this animal, sleeping peacefully at my feet. "Do you want to talk to me?" I asked her. "Would that make you happy?" She raised her head from the floor and gave me one of her classic blank looks: dark eyes fixed on me, her expression curious, calm, and completely inscrutable.

I leaned over to scratch her belly, and I looked into those unreadable eyes, and I smiled. The fact of the matter is, I like not knowing how Lucille experiences the world,

I like the mystery of living with a dog. There is something deeply rewarding about the moments when she and I manage to transcend the language barrier, to reach across the boundaries of species and communicate with one another, understand what the other wants and feels. But there is something equally rewarding about honoring the moments when we can't.

6

OUR DRAMAS, OUR DOGS

Two years ago, Jean, a slight dark-haired woman in her late thirties, was walking down a small side street with her German shepherd dog, then an eleven-month-old pup named Sam, when she saw a man approaching her from down the block. Jean's hand tightened on the leash, and she felt her heart speed up just a bit. Ex-boyfriend. She took a deep breath. Jean hadn't seen this man for almost a year — after a brief fling, he'd broken up with her one night over dinner, made a lot of promises about "staying friends" and "keeping in touch," then promptly disappeared from her orbit — and the prospect of having to stand there on the corner and make small talk with him filled her with dread.

As the man got closer, Sam's hackles rose and he began to growl. This didn't surprise Jean — Sam had been a protective dog since puppyhood, and she was fairly inured to the way he strained at his leash on the street, menacing passersby, so she didn't take the behavior too seriously. She merely whispered to him — "Sam, stop it; it's all right" — and jerked the leash a few times, with little result. But then the man got closer. He waved and called out her name. He opened up his arms to give her a hug. The next few moments are still a blur in Jean's mind. She felt Sam lurch forward, saw a whirl of black and tan fur, heard several simultaneous sounds: barking — deep and guttural and fierce; fabric tearing; swearing. Sam had leaped up on the man, torn a chunk out of his leather jacket, and lacerated his forearm, an injury that required seven stitches.

This is a story about entanglement, about the way an owner's feelings — fears, needs, wishes, both conscious and unconscious — can affect not only our view of dogs but the dog himself and our relationship with him. Jean had acquired Sam nine months earlier following a break-in at her apartment, an incident, she now realizes, that tapped in to a lifetime of fear. "To say that I am an anxious person would be the understatement of the year," she says. The break-in reopened a number of very old wounds — Jean had grown up with an invasive father whose behavior bordered on sexual abuse; at the age of eighteen she had been assaulted in an airport parking garage — and she spent the next several months battling insomnia, walking around with her heart racing. A girlfriend suggested she get a dog, and Sam

— a feisty, intelligent, rather willful pup — seemed like a perfect choice. "I didn't go out and get, you know, a lapdog," she says. "I wanted a big, scary dog. A big *male*. I felt incredibly unsafe in the world."

In retrospect, Jean says she's sure she helped make Sam a big, scary dog. True, like a lot of German shepherd dogs, he had a strong protective streak to begin with — as a young dog, he barked wildly whenever anyone came to the door; he barked at strangers on the street, sat by the window at home, and growled whenever anyone passed by the apartment. But Jean, who found his protectiveness both endearing and reassuring, also encouraged it in subtle, often only semiconscious ways. Sometimes she simply ignored Sam when he barked or growled; sometimes she heard herself talking to him in pleased tones, responding to his menacing outbursts by saying things like, "Are you being a guard dog? Are you protecting me?" In other words, she rewarded the behavior, sent him steady little signals that his aggression was warranted. On a slightly less tangible level, Jean suspects that the dog read her correctly, saw how jumpy she was, felt her tension on the other end of the leash, learned from her to see the world as a fearsome place. And he acted accordingly: as Sam saw it — as Jean helped define it — his job was to perceive danger and fend it off. This was all fine when Sam was young, but a scrappy twenty-five-pound puppy is very different from a seventy-five-pound snarling adult dog. By the time Jean ran into her ex-boyfriend on the street, Sam had become a full-grown menace of a dog who lunged at strangers, growled at anyone who came into the house — friends, the plumber, Jean's seventy-three-year-old mother — and, as it happened, bit people.

Jean says, "You know how people talk about having psycho relationships with animals? They mean me." As the dog became more menacing, her world narrowed. Boarding the dog was unthinkable — Jean couldn't imagine how he'd handle being around strangers twenty-four hours a day — so she ditched vacation plans for a year. Walking the dog became an ordeal, so she began taking him out at odd hours — one A.M., two in the afternoon — when the streets were likely to be empty. Inviting people over became impossible and exhausting, so she whittled down her social life, gradually increasing the distance between herself and others. "My whole life became me and this big dog," she says, and she understands that in an odd way this served a powerful, if only semiconscious, purpose. "Within a year of getting Sam," Jean says, "I was spending almost no time with other humans. It was difficult in a way, but in another way . . . well, talk about *safe*. No one could get near me." Jean had set out in search of a safety net; instead, she'd constructed a fortress.

It is one thing to humanize a dog — to weave little myths about his feelings, to trot out human explanations for his behavior, to see spite where it doesn't exist, or to prepare him a special meal lest he "feel deprived." It is another thing to draw him into more complicated human dramas. Dogs are highly sensitive creatures who read and respond to the cues we give them, even the subtlest ones. As such, they aren't merely objects of human emotion; they can also become unwitting participants in it, and as Jean discovered, the feelings we bring to the bond can affect our relationships with them in striking ways.

Our dramas, our dogs: welcome to the dark side.

In some ways, living with a dog is like being followed around twenty-four hours a day by a mute psychoanalyst: you get that blank screen — nonjudgmental, trusted, noncritical — but no interpretation, no words of insight or guidance, no quiet voice of reason helping you to connect the psychic dots. Feelings float up from inside — rational ones, irrational ones, ones you didn't even know you had — and attach themselves to the dog, who will not question their validity, or hold your behavior up to scrutiny, or challenge your perceptions. Freud in fur; Freud without the therapeutic agenda. In the dog's presence you are free to act — and act out — any way you want.

In their book, *Between Pets and People: The Importance of Animal Companionship*, Alan Beck and Aaron Katcher write about a young woman who went to her vet with lacerations all over her abdomen and thighs, complaining that her dog, a medium-size female, was attacking her. Under questioning, the woman revealed that she slept with the dog at night and clutched her so tightly that the dog struggled to break free, scratching her in the process. The vet told her that this was a perfectly normal response on the dog's part, a reaction to being so tightly confined, and suggested that she give the dog some room. The owner began to weep. She said she couldn't sleep without hugging her dog: she'd been depressed for years, was completely alone and friendless, she needed that level of contact. She then asked the vet to prescribe tranquilizers for the dog, so that she wouldn't struggle against her embrace. When the vet refused, she began to cry again and asked to leave.

This may be an extreme story, and a very sad one for both the human and the animal, but the idea that a dog can unleash so much need doesn't surprise me. Dogs have an uncanny ability to tap in to that level of feeling, a kind of evocative power that helps knock down our defenses and expose sides of the self we tend to keep hidden in human interactions. Dr. Carole E. Fudin, a psychotherapist in Manhattan who specializes in the psychology of veterinary medical practice and relationships between people and animals, thinks that capacity has to do with the canine gaze. "In grief work with patients," she says, "you often hear that the animal was almost a perfect mirror to human emotion: if a client was feeling something intense — sadness or loss — and they looked into the face of their pet, they could not only get a sense of the animal being present to the emotion, but also a feeling of connection, that the animal was truly connected to whatever was going on within the person." Canine sympathy is easy to romanticize, and I don't want to overstate its power. (The first time I wept in front of Lucille, when she was about sixteen weeks old, I looked over at her, hoping to find a little model of empathy — big empathic eyes, an expression of concern; she glanced back at me, then turned, yawned, and calmly began licking her genitals.) And yet I also know what Fudin means. As she's matured, Lucille has become a wonderful solace in the face of tears. If I cry, she sits quietly and watches me; in a gesture that seems to encompass both worry and compassion, she will often place one paw on my arm or knee and just hold it there. This kind of silent sympathy — an ability to sit with someone in pain, to communicate a sense of understanding — is rare among humans, and it helps explain why a person can experience such emotional freedom in a dog's presence.

The language barrier between people and dogs can have an enormously liberating effect, too, inhibiting the internal censors that tend to keep both words and the feelings behind them in check. One day last summer my friend Hope's parents stopped by the dog park to pick up Hope's dog, who was staying with them for a few days while Hope left town on business. When they arrived, Hope's dog raced over, jumped up, and licked them, and I watched all this with complete nonchalance: *Oh, Hope and her folks, isn't that nice.* And then, not two minutes later, I turned and saw Lucille sitting on the grass, quiet and calm and perfectly expressionless, and I heard myself say aloud, "Oh, Lucille, are you sad because you don't have grandparents?" Was *she* sad? I went home that night, vaguely worried that Lucille was "depressed," and it took an entire evening for me to tease out the truth: it always pains me to see people my age, mid-thirties, who still have their parents; it always stirs up flashes of sadness and resentment, reminds me of how suddenly my own parents seemed to vanish from the planet, and while this was a classic example of projection (it was far easier to see sadness in Lucille than to acknowledge it within myself), it was also an example of the way a dog's expression can spark human emotion, touch off feelings we might otherwise keep submerged.

And then there is the dog's profoundly accepting nature, perhaps his most treasured quality. Dogs don't judge us. They are oblivious to the standards humans use to assess one another — appearance and social status, color and class and profession — and so we're often less guarded in their presence, freer to do, say, and feel things we might not do, say, or feel in front of humans. You can roll around on the floor with dogs, sing off key, free associate, and you just don't think about it, just don't think about how stupid or weird you might look or sound. Likewise, you can deflect your sadness onto them, you can communicate to them oceans of fear, you can clutch them while you sleep, and while they may struggle to break free of your embrace, they won't call you on your behavior. In other words, you can be as crazy as you want with the dog, and the dog will never utter a word.

This, of course, is the most wonderfully liberating feature of the relationship — people often talk about being their "truest self" in front of the dog, accepted, unburdened by self-consciousness and fear of judgment — but it can also be a complicating feature, for sometimes that "true" self is not very pretty. And sometimes it can lead you and your dog in some very strange directions.

Example: Lucille is five months old. She is at home, in my bedroom, and I am at the movies, writhing with a degree of anxiety so intense it takes me by surprise. I hate leaving the puppy alone. *Hate* it. Every time

I make a move toward the door, I see that little dog face focused on me, alert and questioning, and I crumble inside. She looks so sad. She looks so alarmed. She . . . I just can't bear it. So I sit in the theater and squirm, check my watch every six minutes. Is she all right? Does she feel abandoned? Does she know I'll be back, or is she terrified?

This is not an uncommon fear: most people I know struggled with guilt and anxiety about leaving their dogs alone as puppies, and virtually all attached dog owners I know admit to melting at the sight of the disappointed dog, the dog who stares after you when you head for the door without him, that picture of dejection. We worry about the dog's sense of time. (Does he know the difference between five minutes and five hours?) We worry about his certainty about our return. (Does he really know what "I'll be back" means?) We are reminded of the domestic dog's essential helplessness each time we leave him alone, of how dependent on us he is. So my anxiety is by no means unique, but the depth of it baffles me. I tend to see myself as a fairly self-sufficient and bounded person, someone who neither needs others too much nor expects others to need too much of me, and all of a sudden — within days of acquiring Lucille — this sense of independence has vanished, and being separated from the dog — even for a two-hour movie — seems like a life-and-death proposition. I check my watch again. I sit. I writhe.

When the movie — interminable — finally ends, I speed home as if heading to an emergency, unlock the door, and race upstairs. Lucille careens toward me, little puppy body jumping up and down in glee, and my relief at the sight of her mixes with a heavy-hearted despair. The room looks like a twister has blown through. Lucille has knocked over a wicker hamper and emptied the contents; the floor is strewn with debris: tattered gym clothes over here, shreds of torn underwear over there, bits of wicker everywhere. She has also pulled down a set of silk curtains: the rod is lying on the floor, and she has eaten a hole in one drape the size of a grapefruit. I sit down on the floor. Oh, Lucille! I don't quite realize this at the time, but the two of us have entered into a very complicated little drama, one part puppy behavior and six parts human emotion. We have become a textbook study in separation anxiety.

Here's what happened, anatomy of a behavior problem: my issue with leaving the dog cropped up within days of acquiring her — just couldn't do it, couldn't leave her alone for five minutes, couldn't *imagine* leaving her alone. The anxiety wasn't entirely unreasonable — Lucille was young and helpless and untrained; I worried she'd be frightened, that she'd pee all over the house, chew up my furniture — but the idea of leaving her alone touched me on a more elusive plane, too, something I couldn't quite identify or understand. All I know is it made me crazy. I'd stand there with my keys in hand, and I'd see her watching me, her whole body a question mark, and I'd want to die. What's happening? Are you leaving me? That's what I'd see in her eyes. But what about me? What will become of me?

Experts give very concrete advice about acclimating a dog to your comings and goings: crate the dog, they say; leave for short periods to get the dog accustomed to your absences; make the departures and arrivals as low key and undramatic as possible so you don't freak the dog out. I read all this, understood all the rationales, and then proceeded to ignore every last suggestion, adopting (to me) a far less stressful tactic: I simply stopped leaving the house. My first six months with Lucille, I don't think I went out to dinner a single time, and if I had to leave her alone for any length, I'd take her to day care, then squeeze everything — lunch, errands, medical appointments — into a single frenzied afternoon. If friends wanted to see me, they came to my house or I brought Lucille to theirs. I took her to all manner of places where she wasn't really welcome — stores, coffee shops, AA meetings — and (more often) I simply wrote off old activities. Shopping? Can't bring the dog. A massage or a manicure? Nope, no dogs. Travel? Are you kidding?

On the rare occasions when I did leave the dog alone, I did so badly. I oozed anxiety. I'd lure Lucille up to my bedroom with a biscuit, and I'd stand there and coo reassurances at her: "I'll be back! I promise!" I'd set up a baby gate to keep her confined to the room, and I'd shoot nervous glances at her, and then I'd turn and flee. Want a behavior problem? This is the recipe. Dogs, after all, are extraordinarily malleable creatures: this is what makes them both so wonderful and also so complicated to live with. As a pack animal, a dog's very survival is dependent on a thorough understanding of his environment — how is the pack operating, who's in charge, what's his place — so he is hardwired to read his owner constantly, to pick up signals about what's appropriate, what's safe, what's dangerous, what's expected of him, what's *happening*. Unlike cats, so characteristically independent and aloof, dogs are highly attuned to their owners, often exquisitely sensitive to shifts in our moods and feelings, and extremely adaptable. That's not to say people don't get emotionally enmeshed and overinvolved with cats — they do — but cats tend to get less involved with *us*: they get up and leave the room if we're bugging them; they don't join us in the social world; they don't absorb and respond to our internal lives in such vivid and direct ways. By contrast, a dog is like a sponge: his instincts are triggered by our moods and behavior, his emotional state shaped by the needs and wants and feelings we communicate to him.

And so it was with Lucille. The more anxious I got about leaving her, the more anxious she got when I started to go, and the harder departures became for both of us. I'd beseech her — "Be good!" and "Please don't eat the pillows!" — and, hearing not the content but the tone and its underlying distress, she'd register only the fact that something very big and scary was taking place. And then I'd leave, and Lucille would do what any self-respecting, distressed pup would do: eat the laundry hamper and chew up six pairs of underwear. I, in turn, would interpret the destruction as evidence of the depth of her distress, and I'd feel even worse about leaving the next time. Problem snow-balled. Vicious cycle ensued. In short order, my internal drama about leaving the house became her internal drama, and I ended up more or less trapped in my own living room, held hostage by a twenty-four-pound puppy.

In truth, Lucille herself wasn't holding me hostage at all; a host of old associations were. One restless night, when she was maybe ten months old, I was lying in bed thinking about how hard it was to leave her alone, and my mind drifted to summers I had spent at a camp on Cape Cod, when I was twelve, thirteen, fourteen years old. I have fond memories of this place — it was a riding camp where each girl got her own horse to take care of, and it's the sort of ex-perience I've categorized internally as rela-tively happy and carefree: nice girls, nice horses, nice setup. But that night, for the first time in years, I remembered that I always spent the first few days of camp bat-tling waves of sadness, a sensation I couldn't explain. The feeling would come up during down times — after a riding lesson, or in the free hours after lunch —

and it rose up like a kind of dread, an anxiety laced with a terrible sense of aloneness, as though I had no safe place in the world. The other girls would be doing things — laughing, reading in their bunk beds, polishing their riding boots — and I'd lie on my bed with my face to the wall, pretending to take a nap and choking back tears. I always called the feeling homesickness, but I think it was something else, a different kind of need. My mother used to drive me to camp, a two-hour ride from our home in Cambridge. She hated driving, and we'd spend long chunks of the ride in silence, her hands gripping the steering wheel. I used to sit there on the vinyl beside her, aware of some vague longing on my part, hating that silence, hoping for something to fill it up.

As I lay in bed recalling this, I realized that what I'd hoped for was reassurance: I wanted my mom to tell me on that drive that she'd miss me while I was gone, to tell me she loved me and would look forward to having me back home at the end of the summer. And then, when we got to camp, I wanted her to hug me good-bye. She didn't, though. My mother was too undemonstrative and shy either to verbalize love or to express it physically — she and my father loved their kids in their own deep and very private way, but they were both markedly reserved people, with an almost pathological aversion to touch — and so when she left me at camp, she'd just give me a shy peck on the cheek and jump back into the car. I'd stand there and watch her drive away, seeds of worry cropping up inside as she disappeared around the bend: Was her reserve a direct reaction to me? Was part of her glad to be rid of me? Would she ever come back? The pain of that is what surfaced when I lay in my bunk at camp, all weepy and forlorn. And the pain of that is what surfaced when I left Lucille. There I'd go, walking out the door without a fuss, without a hug or a kiss or a word of reassurance, and I couldn't believe — just couldn't believe — she didn't experience the same thing I had as a kid.

Such is the power of the dog, that evocative blank screen. If Lucille were a human, she could have looked up and said, "Hey, don't get tied into such a knot about leaving, I know you're only going to the movies — I'll be fine." Because she's not, I read into her eyes what I felt: *But what about me? What will become of me?* Those, of course, are *my* questions, the big dark existential questions I've lugged around my whole life, and in the inadvertent, wordless way of a dog, Lucille has slowly circled me back to them, back to the central insecurity I felt as a kid. This is ancient stuff, territory I've stamped across over years of psychotherapy, but when I'd leave the house and see her worried expression, it all got reflected back, that whole history of fear and doubt. Those eyes: I'm twelve years old, waiting for my parents to come home from a party, certain they've died in a car wreck and won't return. That look of alarm: I'm thirteen, watching as my father swims in the ocean off Martha's Vineyard; he's a tiny dot of a man in the sea, and I can't take my eyes off him, can't lose sight of him, because if my vigilance wavers for even a fraction of a second, he will surely drown, disappear out there before I've mustered up the courage to tell him I love him. That anxious gaze: I'm six or seven, I'm ten or fifteen or even thirty-five, and I don't know — I just don't know — if the people I love love me back, if they can give me the things I need.

Shortly after I made this connection, I bit the bullet and started taking the experts' advice, making my arrivals and departures far less emotional, handing Lucille a bone smeared inside with a little peanut butter when I left the house so she'd associate my leaving with something positive, that kind of thing. She's long since stopped tearing stuff up, she doesn't howl or whine while I'm away, and when I return, nothing in her manner suggests she's been the least bit anxious or stressed. As far as I can tell, she does what most dogs do when left alone: curls up on the sofa and sleeps. So I can leave her alone these days, and I do leave her alone, but I'm still loath to, and I'm still struck by the range of feeling that that look in her eyes can generate in me.

If there were a 12-step program for codependent dog owners, I'd no doubt be a prime candidate, but I'd hardly be the only member. Right there in the church basement with me would be Jonathan, forty-one, owner of a four-year-old basenji named Toby, and a man whose separation anxiety matched mine, step for angst-ridden step. Jonathan barely let the dog out of his sight for an entire year, and it wasn't unusual for friends to find him at work in his office, the dog curled in his lap. Next to Jonathan would be Elizabeth, thirty-nine, who worries so excessively about the health and safety of her dog, a beagle named Marge, that she falls apart if the dog so much as hiccups. And beside her would be Joan, thirty-four, who's gone on vacation without

Emma, her springer spaniel, only once in three years, and who felt so guilty about that departure that in the days before her trip, she showered the dog with treats — rawhide, marrow bones, an abundance of table scraps. By the time Joan left, Emma looked miserable and lethargic and wouldn't eat her dinner. Joan assumed she was depressed about her impending departure; in fact, she'd developed some kind of painful gastrointestinal block and had to be rushed from the kennel to the vet while Joan was away.

Take one human struggle, add dog, and stir: recipe for entanglement. Jonathan, executive director of a group of addiction treatment centers in Greater Boston, has a long history of overinvolvement in human relationships, of blurred boundaries and excessive caretaking, and his codependence came out in spades with Toby. "The problem for me was not being able to control what happened when I wasn't there," he says. "And for someone who wants to be in control of just about every part of their life, leaving this animal alone, who's so dependent, was horrible." The horror was compounded by the circumstances under which Jonathan acquired Toby: he inherited the dog from his lover, Kevin, who died of AIDS, and few emotions can fuel one's need for control — and one's fear of losing it — like profound grief. Jonathan's fear, his need to manage the people and things around him, went into overdrive with Toby, who, much like Lucille, read Jonathan's anxiety and responded in kind. Toby is a small and delicate dog, with a sleek tan coat and tight curl of a tail, but he could tear through a room in about ten seconds. Same vicious cycle, similar outcome. Jonathan took Toby everywhere — to work, to meetings — and the aspects of his life that couldn't accommodate the dog more or less ground to a halt; by the end of a year, people at work were taking Jonathan aside and saying, "You know, I'm really not sure it's appropriate for you to be bringing Toby to this meeting with our major funders." Jonathan would bring him anyway. Time has eased the struggle (so has a crate), and Jonathan can laugh about it today, but his story is a textbook study of codependence between owner and dog, in all its deflected pain.

Human struggles don't always affect canine behavior so directly, but the emotions they generate can drive owners halfway 'round the bend. Elizabeth, for example, can't count the number of nights she's spent staring at the ceiling worrying about her dog's death, *weeping* over it. She knows this is crazy. Her dog is only four years old; she's a perfectly sturdy little beagle, thirty-five pounds of perfect health, but Elizabeth can't help it. Her mother died of cancer when Elizabeth was fourteen, her grandmother died three years later, and her older sister two years after that. Elizabeth's grief has abated over the years, but her sense that the world is dangerous — that the people you love can be snatched up at any time — has not, and the fear she carries around with her gets channeled into Marge, translated into a powerful need to protect and watch over her. Elizabeth herself hasn't been to a doctor for so much as a routine checkup in years, and she's the sort of person who'll barely take an aspirin when she has a fever or a bad headache, but if she sees anything amiss with the dog — a limp, an untouched meal, a look of lethargy — she's all over her: What's wrong with Marge? Is she sick? Is she *dying?* "I don't feel any of that for myself," she says. "I don't worry about my own safety or health, I don't ever worry that I'm in danger — it all goes onto Marge, and it's amazing how much emotional energy I can expend worrying about her."

Joan's emotional energy goes into fending off guilt, the bane of many a dog owner's existence. Everything makes Joan feel guilty — leaving the house for work in the morning, cutting a walk short, expressing even slight irritation — and she has found the power of this emotion both oppressive and illuminating. Joan describes standing at her back door one cold winter morning imploring Emma, who was out in the yard, to come inside. She was in a rush, overtired, premenstrual — bad combination all around — and the dog wouldn't come, and Joan lost her temper. Clad in slippers and a robe, she went into the yard, grabbed Emma by the collar, and dragged her inside, jerking her by the collar and hissing. Then she got ready for work, left the house in a huff, and spent the entire day writhing with guilt at her desk: she'd made Emma feel bad, made her feel unloved and inadequate somehow. Not the most dramatic story — and not the most rational set of interpretations — but it taught Joan something about her own hypersensitivity to the dog's feelings. She says, "I keep assuming that the dog feels exactly the way I used to feel with my mother, who was this extremely perfectionistic woman who always made me feel less-than. 'I'm a disappointment, nothing I do is good enough, and one of these days I'm going to blow it. She's going to stop loving me.' I grew up with that feeling. So I guess I feel like the stakes are always incredibly high. If I let Emma down in any way — you know, if I yell at her or make her sad — it's like she might feel all of that, too, like she's this major disappointment." Joan's heavy hand with the pre-vacation treats was a wake-up call, showing her just how out of proportion her guilt and anxiety had become, and how dangerous those emotions could be.

Gathered there in a 12-step meeting for dog owners — let's call it Ala-Pet — none of us would be particularly proud of the way our own emotions tend to get tangled up with our dogs, and all of us would agree that it's dangerous business, that things can get pretty complicated when you start acting out psychic dramas with an animal. You can create excessive dependence, compromise a dog's physical health, make yourself half crazy with worry. You can also do far worse things, and many owners do — animal abuse is perhaps the most vivid expression of the way human emotions (violence, hostility) can harm dogs. But in the lexicon of 12-step programs, all of us would admit to a certain sense of powerlessness, a periodic and unwitting loss of reason in the face of the dog. As Elizabeth puts it, "They just touch you on that level, I don't know how else to explain it. You can be an incredibly rational person in most areas of your life, and then with the dog you're a total lunatic." Amen.

Mordecai Siegel, author of fifteen books about dogs and current president of the Dog Writers' Association of America, once said, "Acquiring a dog may be the only opportunity a human has to choose a relative." That's a lovely idea, but human relatives — even self-selected ones, such as spouses — don't tend to be as accepting of our foibles as dogs are, and they're not nearly as pliable. Humans can't be trained, and they rarely give their partners so much emotional free rein, rarely permit them to act out with such unscrutinized abandon. So I'd alter Siegel's sentiment slightly and say that acquiring a dog may be the only opportunity a human has to *create* a relative, to make up the rules as you go along, to define the terms, all to your own specifications.

There is good news and bad news in this state of affairs. The good news is that dogs are very adept at molding themselves to our wishes. They are such accommodating animals, equipped with such keen radar, and they really will play by our rules if we make them sufficiently clear: you want minimal affection from the dog, he'll learn to keep his distance; you want the dog to occupy a secondary role in the household, he'll take it; you want to give him an hour of exercise every day, or twenty minutes, or two hours, he'll learn to expect that amount, no less and no more. This can make for a very gratifying partnership indeed: dogs represent the one relationship in life you really can "design" to some extent, and because dogs respond well to both structure and hierarchy, the result can be satisfying to both parties. The bad news is that humans can be woefully fuzzy on the concept of dominion, of the responsibility it entails: we don't always understand canine behavior very well, we're not always very clear about what we want from dogs or what we're asking from them, and our comparative lack of sensitivity can lead both parties — dog and owner — into trouble.

A small but telling example: not long ago, on my way out of the house with the dog, I discovered that the lock on the front door was jammed. Couldn't get the key in, couldn't get the key out, it was stuck at the halfway point. I was in a bad mood and also in a hurry, so this annoyed me no end. I slammed down my bag, dragged the dog back inside, stormed into the kitchen muttering, rubbed some olive oil on the key, then went back to the front door. Bingo: key slid in, problem was solved.

Lucille was not aware of this, however. Lucille was six or seven steps behind me in this process, still fixated on the fact that I was angry. So when I finally locked the door and turned around, there she was, an expression of the gravest concern in her eyes. She looked up at me imploringly, then leaped up and placed her front paws on my arms. Her ears were flat back and her tail was wagging madly; she curled her upper lip back, baring her teeth in a submissive smile, and craned her neck up to lick my face. She seemed to be saying, *Please don't be mad, I didn't do anything bad, I swear.* My heart melted at this, but I was also struck by that quality of canine attunement: how keyed in to us dogs can be, how responsible for us they seem to feel, and how inadvertently our moods and behaviors can affect them. I looked at her and thought: God, it would be easy to screw this animal up. A dog's sensitivity is an exquisite thing, but it can be dangerous in human hands, for we're not always aware of what we're communicating to them, of what needs and feelings might be simmering below the surface.

Consider a woman I'll call Martha, who became charged with the care of her six-year-old mixed-breed dog after her husband died. Martha's husband was a nasty man — alcoholic, abusive — and on a fairly regular basis he'd get drunk, pick a fight with Martha, then corner her on one side of the living room and slap her. During a visit to her vet — a routine call, for the dog's rabies vaccination — Martha confided that she was having "some trouble" with the dog, that he didn't seem to like her. When the vet questioned her about this, Martha let on that following her husband's death, the dog had begun to charge over to her as she passed through the living room, corner her against the wall, and snap at her, baring his teeth and growling.

The vet who reported this story speculated that the dog's aggression may have been a fairly straightforward example of learned behavior — taking his cues from his former master, the dog determined that, for unknown reasons, Martha needed to be cornered and snapped at; since the husband's death, that job now fell to him. But he also speculated that Martha may have played a role in the confrontations, too, that these daily skirmishes reflected learned behavior on *her* part as well, that she was unwittingly reenacting with the dog the drama she'd played out in her marriage, sending him signals that the aggression was somehow necessary and appropriate — a kind of human-dog version of battered-wife syndrome. Blaming the victim? Perhaps, but I'm enough of a believer in the power of the unconscious to think there may be some truth to that explanation, to the idea that the dog picked up on some signal of Martha's and acted accordingly.

Certainly I believe that dogs absorb our worldviews, that they're very adept at figuring out who we want them to be and sliding right into the role. Bonds between people and dogs are like miniature marriages, sometimes enormously successful and healthy, sometimes dysfunctional and weird, but always based on some complex blending of need and temperament, communication and miscommunication.

My friend Joanne complains that her mother has a "psychotic" relationship with her dog. The mother, according to Joanne, is an angry, embittered, rather paranoid woman, a widow in her late sixties, and the dog, a four-year-old terrier mix, is apparently an angry, embittered, rather paranoid animal. The dog is dominant and highly territorial; he snarls at other dogs on the street, barks furiously whenever anyone comes to the door, and has bitten a number of visitors, including Joanne's husband, twice. Joanne keeps telling her mother to do something — the dog is out of control, she needs to take him to obedience classes, *there is a problem here* — but the mother refuses. When the dog bit Joanne's husband, the mother defended the dog: Alan must have provoked him, she said; it couldn't have been the dog's fault.

Not long ago, while Joanne was walking with her mother and the dog, a young woman with a miniature poodle passed by, and the two dogs started snarling at each other, a little dance of distress and aggression. Joanne's mother picked up her terrier and held him to her breast, then started screaming at the other woman about her "vicious dog." When the woman was out of earshot, she put the terrier down and started railing to Joanne, a venomous monologue about people, about the world, about how you can't go anywhere these days without running into trouble. And it dawned on Joanne: the dog is her mother's perfect partner, as scrappy and territorial and angry as she is. She's encouraged that behavior in him — reinforcing his fearfulness of other dogs, for example, by picking him up in horror; failing to take charge when he acts aggressively — and she also needs it. The dog is the repository for all her mother's feelings of ill will, all her dark, angry pessimism; he acts those feelings out for her, gives voice to her own fear and hostility.

A caveat: I don't believe that most people enter into complicated emotional entanglements with their dogs, and I don't believe that every dog owner in America is unconsciously foisting his or her deepest fears or neuroses onto the dog, leading the animal into some dark subterranean drama. A lot of the strong feelings that come up around dogs — guilt about leaving them alone, worry about their health, anxiety about their well-being — are simply products of attachment, pieces of the emotional landscape we inhabit when we love another being deeply. But I also think the potential for psychic drama is high with dogs. They are, after all, among our most private relationships. We spend most of our time with dogs behind closed doors, in our own homes, where we're most likely to be ourselves. The larger world rarely observes us in close interaction, so the dynamics between us aren't subject to the same degree of social scrutiny that characterizes relationships between humans. (If something weird is going on between you and your dog, you're not going to get a call from his first-grade teacher, asking why he's stressed out or depressed.) Nor are relationships between people and dogs often subject to the same degree of *personal* scrutiny as human relationships. Because dogs don't contradict or negotiate with us, because they tend to act on cues so subtle we're often unaware we're sending them out, we can be blissfully unaware of what's going on with dogs in the first place, what we're asking from them, how we're using them or interpreting them or shaping our alliance with them. Dances between people and dogs are choreographed in confidence, moved by music so subtly scored, we may not even be conscious of its sound. We feel, we move, the dog moves along with us.

If we're lucky — and wise enough to learn from our mistakes — these dances can actually have a kind of elegance, taking us places we need to go. Sometimes they can help us cope with feelings that are hard to deal with in the human world, providing a relatively safe way to act out some of our more complicated impulses. "If I didn't have Marge," says Elizabeth, "I don't know what I'd do with all that fear. Maybe I'd be more of a hypochondriac about myself, or about my friends, I don't know. But there's some way that she contains the fear, like it's all housed right there in that little beagle body, where I can sort of deal with it." Jonathan agrees: because Toby so satisfies his need to care for and manage another being, he now feels less compelled to act on those impulses with humans. "I think that in a lot of my early relationships," he says, "I was treating people kind of like pets — trying to control them, not being able to give them space. Toby is like a big release for those feelings." He's not a bad release, either: Toby is a dependent creature, who by definition requires a certain degree of caretaking and control. "I sometimes think a lot of my human relationships would have been better if I'd had a dog twenty years ago," Jonathan says. "For a truly codependent person, a dog is like a dream come true."

Entanglements with dogs can also serve more complex purposes. In retrospect, Jean says there was "a weird kind of beauty" about the dance she entered into with Sam: she couldn't have dreamed up more compelling choreography if she'd tried. Not only did the dog inadvertently help her withdraw from the world, establishing a canine wall between herself and others; he also allowed her to recreate something familiar — a world in which her life was dominated by a large threatening male. But this time Jean added a twist, rewrote the script: this time the male was threatening other people, not her.

"There was something very satisfying about living with this big scary dog," she says. "In one way, I felt as powerless and frightened as ever — you know, my heart froze every time the doorbell rang, because I had no idea how Sam would react. But in another way, I felt completely ensconced in there with him: no one could get me. Literally." Friends told Jean she had to deal more effectively with the dog — get him under control, take him to a behaviorist or a trainer, do *something.* Ultimately, after the episode with her ex-boyfriend, she did see a behaviorist and enrolled in an obedience class, steps that helped her communicate with Sam more effectively, elevate her place in the hierarchy, tell him when his protectiveness was warranted and when it wasn't. But she resisted this for a long time, almost a year, and she understands why. "A part of me knew I'd created this big huge problem — I'd helped create this aggressive, hypervigilant beast — but another part of me just wanted to stay exactly where I was with him. He wasn't just a dog to me: he was my weapon in an unsafe world."

I suppose the same could be said of Lucille, nonthreatening though she may be. Get close enough to them, and dogs — like humans — can tap in to several different dramas, pulling separate heartstrings simultaneously. I can see, for example, that the dance I entered into with Lucille over leaving the house spoke not just to the drama of my early childhood but also to more current struggles: with intimacy, with my own fears about the human world.

Sometimes when I come home with the dog after our final outing of the day, when I shut the door and lock it behind us, I'm aware of a profound sense of relief, a sense that we've shut ourselves off from the scary world out there, that we (or more to the point, I) can relax, let the defenses down, breathe. Those moments sometimes seem to save me, so pure is the sense of security behind them. *I'm done.* That's the feeling. *I've done everything out in the world I have to do, I'm at home alone with my dog, I don't have to be scared anymore.*

The fear I'm referring to, the fear I've just locked out behind me, can stem from the simplest things — professional engagements, social commitments, even errands — all of which fall under the larger category of human interaction. And human interaction can be scary stuff, especially when you're newly sober and moving through the social world with no access to anesthesia, unable to dodge your central feelings of disconnection and uncertainty. *I'm done. The scary world — the place where people disappoint and confuse you and even die — is out there, and we are in here, safe at last.* That feeling appeals to me as viscerally as Jean's sensation of being "ensconced" in her apartment with Sam, her four-legged amulet. Lucille has been an excuse for it in a sense, providing the rationale I've needed to turn my own home into a hideout. It's easier to say "I can't go out" than it is to admit "I don't want to." Easier to say that *she'll* be afraid than to admit how fearful I am.

This strategy, of course, has its flaws. Yielding to the wish to run and hide can be isolating. It can affect the other people in your life, who get shut out along with the rest of the world when you lock that door behind you. It can, in a word, create boyfriend trouble.

7

FAMILY DOG

Scene: a morning in mid-November, about fifteen months after I've gotten Lucille. My boyfriend and I are sitting in a couples therapist's office. I am speaking, near tears. This is our first meeting with the psychologist. We are talking about . . . the dog.

"I want to feel like she's *mine*," I say.

"But she *is* yours," Michael says. "The dog adores you."

"But . . . but . . ." I choke, half-formed thoughts about love and trust and exclusivity trapped somewhere between gut and mouth.

Lucille, it is safe to say, was an "issue" in our relationship from the beginning. This sounds ridiculous, like something you'd hear on a daytime talk show ("Women Who Love Dogs Too Much"), but it was true.

From the day I got her, I was a total hog with Lucille. Mine, mine, mine. The dog is *mine*. Pre-Lucille, I spent four, five, sometimes six nights a week at Michael's house. Post-Lucille, I started to spend three nights there, maybe only two, and I was starting to feel tense at even that number, compelled to be back in my own home. I had a rationale for this: my house has an enclosed patio, so when Lucille was a puppy, I could take her out to pee in the middle of the night without having to get dressed and put on shoes, whereas Michael lived in an apartment, with no access to fenced-in space. It was therefore more practical to stay at my house. But in truth I wanted the dog to myself. I wanted her to bond with me and me alone, and the ferocity of this possessiveness took me completely by surprise. I wanted her to follow me and not him, to sleep on my side of the bed, not his. If we were all sitting on the sofa and she put her head on his leg or curled up against him, I'd get a horrible, mean-spirited little stab of jealousy, and I found this so painful and embarrassing I couldn't even talk about it. Instead I started angling for more time alone. "I could use a night to myself," I'd say. Or, "I think I'll stay at my house tonight," and neither Michael nor I chose to comment on the fact that I didn't ask him to stay there with us. This made me feel horribly small and mean and tense, all this orchestrating of distance, but I couldn't help it; the reaction was so visceral it overpowered me.

Michael is probably the nicest man I've ever known, and by the time we started seeing the couples therapist, I'd known him for seven years. He'd been my primary caretaker all that time, and without question my best friend. I met him just after I'd stopped living with my old boyfriend, who was not a nice man at all, and Michael literally held my hand through that breakup, which dragged on for several years. I remember calling him up from work one day, just after I'd left Julian and moved into a new apartment, and weeping into the phone, telling him I thought I was having a nervous breakdown. He said he'd meet me for a walk in the Boston Common, and we sat on a bench in the sun. I cried and cried and talked about how miserable I felt about the breakup with Julian, and Michael listened, his arm around my shoulders. That's how he always was: a man who'd listen and hold you even when you talked about things that should, by all accounts, have hurt or dismayed or warned him away. Sometimes I thought this was a sign of deep generosity, and sometimes I saw it as an inability on his part to set limits, but whatever the motive, Michael is nothing if not steadfast. He saw me through the eleven months my father was dying, and a year later he saw me through the death of my mother, and eight months after that he saw me through my decision to quit drinking and go to rehab. He cooked a million homey dinners for me through that time, rigatoni with red sauce, and chicken with dumplings, and Italian sausages with mashed potatoes, and he almost never called me on the fact that I didn't give nearly enough back.

I got Lucille without consulting him. I'd told him the morning I picked her out that I was just going to *look* at the shelter, and later that day, after I'd taken her home, I went over to his house to show him. Lucille trotted in, looking edgy and anxious, and peed on his carpet within thirty seconds. Then in short order she defecated twice, once in his living room and once in the bedroom. In retrospect, this seemed oddly apt to me: it was as though Lucille were delivering a little message from me, making a statement about how much of a mess I could make of things. Michael was annoyed but characteristically noncombative about this: I took Lucille outside, he cleaned up the mess, and he never called me on the fact that once again I'd gone and made a big life decision without him.

I'd done this with my house the year before, deciding almost overnight to buy a place in Cambridge that I knew was too small to accommodate both of us. I said it might: at some ill-defined point in the future, I said, we could turn the third floor into a work space for him, or maybe we could build an addition off the kitchen, but inside, I think I knew: *My* house. My space. Not ours.

Same with Lucille. *My dog. She's mine.* I kept hoping this would ease, kept hoping I'd relax about her a little, allow Michael to share in the caretaking and responsibility and delight of her, but that fierce sense of possessiveness persisted and persisted, and I simply couldn't let him in. I hated this about myself, hated feeling that selfishness rise up, but like I said, I couldn't help it.

Dog as symbol, dog as mirror, dog as barometer of human affairs. We tend to think of dogs as sweet and easy adjuncts to family life, simple beings with simple roles: the dog doles out affection to the nuclear unit, the dog offers the kids companionship and lessons in responsibility, the dog protects the family home. Dogs can — and often do — perform all those functions, but they often execute other tasks, as well, reflecting — and sometimes participating in — much more complicated aspects of family life.

Lucille turned out to be an expression of my limits with Michael, my inability to share my most important stuff. About a week after I got her, Michael and I were driving in the car with Lucille, and I made some reference to him as the dog's uncle: Uncle Michael. Michael turned to me and said, very definitively, "Uncle, nothing. Uh-uh. I'm *Dad.*" That jarred me, the insistence in his voice, and I didn't say anything, but inside I was thinking: *Nope, I'm sorry. You're the uncle.* For a long time after I got her — for a good year — Michael would talk about us as a pack: Lucille seemed happiest, he said, when the three of us were together, when the pack was reunited. This made me feel unbearably guilty and conflicted, the hope behind such statements, because I couldn't share it, couldn't reciprocate. In my heart, Lucille and I were the pack, that pack of two, and Michael stood just outside the circle, close enough to be near it but a safe distance away from the center.

This feeling wasn't just a by-product of ambivalence toward Michael, or toward the idea of making a deeper commitment to him, both of which were realities that predated Lucille by some time. Instead, it was driven by a feeling of need that may have had very little to do with him: I *need* this, I need this dog to myself. That was the sensation: I need to cultivate a sense of belonging and attachment to this dog, and I need to do it alone, in order to learn that I'm capable of it. I need to love her, and to have her love me, before I can expand the circle, complicate it any further. The selfishness that sprang out of that need — the sense that I couldn't allow Michael to share in the bond or attachment — made me feel guilty and mean, but in some ways I was like a kid who's been denied candy for a long long time and then goes berserk on Halloween, grabbing treats by the fistful, then guarding the stash. *I need her all to myself; I'm so hungry for what she can give me.* This was a variation on the same theme that cropped up when I had to leave the dog alone, a feeling, born in childhood, that love was somehow terribly fragile and tenuous, a limited resource that needed to be constantly protected and constantly reinforced. *She's mine. I can't leave her, and she can't leave me.*

This saddens me, to think about how unwilling I was to share the dog, but I think the phenomenon is not uncommon: get a dog, and whatever strengths and limits characterize a relationship — levels of commitment, degrees of competitiveness, areas of conflict — can be highlighted and underscored. A dog can create alliances or cause them to shift, illuminate difficulties or help

223

mask them, expose the inner workings of the existing pack faster than you can say, "Rover, no!"

Add dog and stir: sometimes you get a family. A lot of couples I know have gotten dogs and ended up with more solid bonds themselves, this pack-oriented creature in their midst helping them to see themselves as a family, or as potential parents, deepening the sense of commitment. My friends Beth and David adopted a two-year-old German shepherd–Siberian husky mix, got married within a year, had a baby a year later. The dog made them realize how much love they had to give; sharing in her care made them recognize each other's skills as nurturers. Same thing happened to a woman I know in California, also named Beth. She adopted a shepherd mix about a year before I got Lucille, ended up moving to San Francisco with her boyfriend Andy, got married last summer. The dog cemented the relationship, helped turn a pack of two into a pack of three.

Add dog and stir: sometimes you get a disaster. "I think dogs make people break up," says Liz, a journalist who lives in Chicago. Her experience had to do with degrees of attachment: her boyfriend got her a puppy for Christmas, she fell madly in love with the dog, then realized that same depth of feeling didn't quite extend to the boyfriend; she didn't feel nearly as devoted to him, nearly as charmed or committed, she wanted the puppy all to herself. Six months later the boyfriend was history.

Jessica's boyfriend was history, too: they got a puppy together, the boyfriend turned out to have a far more punitive and physical approach to discipline than she did, and any fantasies Jessica may have harbored about him as a potential father went straight out the window. Angry man, mean to the puppy, easy equation: keep the puppy and dump the man. It is axiomatic that you can learn volumes about people by watching them interact with a dog, seeing how much kindness or affection or playfulness the dog evokes; it is also axiomatic that you can learn volumes about a third party by watching your dog's response, and some of us use this principle to great effect. "A girlfriend of mine used to live downstairs from us," says Wendy, who works in public television in Los Angeles, "and every time she had a date, my husband and I would send the dog down to check the guy out." The dog in question is a tiny white Maltese named Baci, who is equipped with marvelous boyfriend radar, and Wendy's friend wouldn't form her own impression about a prospective man until she'd gauged his reaction: a few good wags usually led to a few good dates; if the dog shied away from the guy, or growled or seemed disturbed, she'd write him off. Dog as soul-sniffer: a handy skill.

Baci served an important function on the domestic front, too. "I know all these working couples who are thinking about starting families," says Wendy, herself part of a dual-career couple, "and they're just starting to address the big questions about responsibility: who's going to take care of the kid, who's going to cut their hours back. I know having a dog isn't the same as a kid, but my husband and I have been having those conversations for years." Which one of them will feed Baci; which one of them will take time off to bring him to the vet; who's doing too much or too little: "Dogs," Wendy says, "really do force you to work that stuff out."

Before they got a dog, my friends Polly and Wendy had to sit down and have a conversation they'd been putting off for eons, a long dialogue about a seemingly trivial, but actually loaded, subject: housework. The dog would add more items to their daily list of Things to Do, so questions arose: who would walk the dog, who would take the dog out to pee at night, who would brush and groom the dog, and — in the midst of all that — who would clean the house, cook dinner, see to the grocery shopping, and so on? Wendy is thirteen years older than Polly, she owns both the condominium they live in and most of its furnishings, and she'd always felt as though she bore too much of the responsibility for long-term household care: getting the rugs cleaned, for example, and making sure the furniture got polished. Polly, on the other hand, felt she shouldered too much of the physical work, the heavy lifting, like taking the trash out. Neither felt the balance was right, and getting the dog forced them to look more closely — and talk more openly — about the division of labor in their household, and at the old anxieties and tensions that had lurked behind it, un-addressed, for years.

Their story had a happy ending, but dogs don't always lead couples toward resolution. Carolyn and Mark, a science writer and a carpenter, had been married for nine years, seven of them rocky, when they decided to get a puppy. Their motives seemed benign enough — they're both great dog lovers — but Carolyn suspects they each had a hidden agenda, much like a troubled couple who hopes a baby will help patch up a marriage. "We didn't frame it that way," she says, "but I think we both hoped the dog would be a positive influence, that it would give us something to share and take care of together, something we wouldn't fight about."

Perhaps not surprisingly, the dog — an extremely high-energy, high-maintenance weimaraner named Jojo — had the opposite effect, bringing to a boil issues that had simmered between them for nearly a decade. Carolyn, now in her late thirties, is a woman who takes responsibilities very seriously — balances the checkbook down to the penny, makes sure the car gets tuned up regularly, worries about things like chipped house-paint; her husband, Mark, had a far more laissez-faire approach to life (and to Carolyn's mind, a far less mature one): chronically late, never paid bills on time, did his share of the household chores reluctantly, sporadically, and inefficiently. They clashed constantly over Jojo. Carolyn took him religiously to obedience classes; Mark promptly undermined everything she taught him. Carolyn didn't want to let the dog off-leash until she was certain he'd come on command; Mark would take him out and let him run around like a maniac.

They fought about feeding the dog from the table, letting the dog on the furniture, neutering the dog; they fought about using a choke collar, they fought about diet and exercise, they even fought about clipping Jojo's nails. (Carolyn thought it was important; Mark thought it was "stupid" and refused to assist her.) Above all, they fought about who shouldered more responsibility.

The situation became intolerable to Carolyn: the poor dog, receiving inconsistent and mixed messages from his owners, became increasingly out of control — "a sixty-pound wildman" — and her view of her marriage became increasingly clear. "The dog brought out so much conflict. All his immaturity. All my resentment. All our differences. And mostly this feeling I'd had for years: that I was living with a man who just couldn't be counted on." Within a year of acquiring Jojo, Carolyn and Mark separated, then divorced. Subject of their last major blowout: who'd keep the dog. (Big sigh of relief: Carolyn won.)

Carolyn says a lot of her friends think she and Mark broke up solely because of Jojo. "I think people thought it was pretty weird," she says. " 'You left him because of the *dog?* Who breaks up a marriage over a *dog?*' But I don't think non–dog people understand how much *stuff* can come up over an animal."

Oh, yes, buckets of stuff, great heaps of it, in many different forms. Dogs tend to home in on one member of the family, to single out one person as *it* — Alpha God, king of the universe — so they can inadvertently cause jealousies to crop up, private insecurities and dormant feelings of competitiveness. A woman I know named Sue cannot stand the fact that her dog, a collie mix whom she feeds and walks and cleans up after and *adores*, will literally throw herself at her boyfriend's feet when he walks in the door. Cannot stand it. Sue does all this work — hours a day with the dog, volumes of emotional energy expended on the dog — and if Matt is in the room, she feels like she just disappears from the dog's radar.

Dogs, it turns out, know nothing about human sexual politics, and this is one of the sad but true realities (some might say annoyances) of living with a dog, particularly a female — some of them can be virtual whirlwinds of submission in the presence of men: they flirt, they roll over at the merest whiff of testosterone, they drive their good hardworking, feminist owners crazy in the process. Like a lot of women, Sue, a lawyer, forty-one, has spent the bulk of her adult life trying to develop and come to terms with her own strength, struggling to be independent and self-reliant and proud of her own competence. Seeing her own dog writhe in submission at her boyfriend's feet feels enormously undermining to her, as though the dog is acting out some piece of her own self she's spent years trying to reject. Her reaction? She gets mad at the boyfriend, territorial with the dog. The first year she and Matt were together, all their fights were about the dog, Sue trotting out laundry lists of complaints that look petty to her today (Matt was being too indulgent with the dog, or too playful; Matt was undermining her training and her authority) but that actually spoke to much deeper tensions. The secret refrain in her head: The dog loves (respects, admires, responds to) him more than she loves me.

You hear the opposite refrain among couples, too: You love that dog more than you love me. It is a fact of family life that dogs tend to be a central focal point for affection — often for more of it than the human family members. In a pioneering study of the role pets play in family systems, psychiatric nurse Ann Cain analyzed degrees and frequency of "stroking" (defined as any form of positive recognition, such as touch, reassuring smiles, or gestures) among sixty families with dogs: 44 percent of the families surveyed said their dogs got most of the strokes in the family; only 18 percent said dogs and family members got an equal number of strokes. This may not be a surprising finding — it's easy to focus attention and affection on an animal; loving words and gestures aren't emotionally loaded with dogs the way they are with people — but it's not hard to see how the phenomenon can cause old tensions to bubble up. Husbands complain that wives dote on the dog, baby the dog, lavish the dog with affection; wives complain that husbands are "nicer" to the dog than they are to them or to their kids.

Dog as focal point. My parents were among the least verbally expressive people I've known: emotions ran deep in our household but were rarely expressed openly. No fighting, no raised voices, no big emotional displays of any kind, and this was true in their dealings both with each other and with their kids. The one exception: the dog. I can remember long, somber family dinners at my parents' house in Cambridge, the five of us sitting in a characteristic strained silence, light from candles flickering on the table, no one saying a word — and then a periodic explosion as my father yelled at the dog for begging for scraps. He'd bellow at the dog — "Tom, go lie down!" — and the dog would slink off to the other side of the room. We'd all swallow hard and return to our silence. Times like that, you got the feeling that a leak had sprung in the dam somewhere, that whole rivers of rage ran beneath my parents' quiet reserve, that outlets for it were painfully few and far between. The dog provided one outlet, perhaps the safest one they had.

That applied to more positive feelings as well. The strained family dinner was invariably preceded by the strained family cocktail hour, my parents sitting on the sofa in the living room with their drinks, the dog lying on the floor in eternal hope: Might someone drop a peanut on the rug? And if so, could he make a move for it? If a nut dropped — even half a nut — the dog would pounce, with such ferocity you'd think a squirrel had just darted out from under the coffee table. Watching this always made us smile, the glint of victory in his eyes when he scored the errant peanut. Toby, their second elkhound, had a particular fondness for martinis: he'd come up to me or my father, give the glass a good sniff, then step a few feet back, sneeze wildly, and come back for more. As a family, we didn't laugh a whole lot among ourselves, but we always laughed at the dogs: they were a safe topic of conversation, a ready source of amusement, a tension reliever.

About a year after she divorced her husband, Gina, forty-seven, came home to find her two sons, then ages ten and twelve, with a stray dog they'd found on the street. Ugliest dog you've ever seen, Gina says, mangy and coyotelike, with matted hair and bald patches in his coat, but they took him in, named him Lucky, and he very gradually became a symbol of the family's new life, in the aftermath of the divorce. "This dog did not know how to play," Gina says. The boys would throw him a ball, and he'd just stand there, as if to say, Well, if you want the ball, why did you throw it away? They'd wave a chew toy in his face, and he'd back away, suspicious and confused. Within a year the kids had taught the dog to play fetch and Frisbee, they'd developed an elaborate game of tag, complete with rules that the dog understood, and they'd taught him a number of skills to execute indoors: Lucky could turn the TV on and off with the remote; he could balance a biscuit on his nose; he could smile on command, baring his teeth when they called out "Cheese!" After the divorce, Gina says, "I think we'd all sort of forgotten how to have fun, and there's a way in which Lucky gave us a new way to play together, a focus for it: the dog in the backyard instead of the dad." She pauses, then adds, "Also, the dog is much better at making us laugh than the dad ever was."

It's not uncommon to hear people — particularly women — talk about the connective effect of a dog, often to a remote or unapproachable father. "Growing up, the only person my father could relate to was the family dog," says Nancy, a New York interior designer. "He was a completely isolated person, and none of us was close to him, but because he and I both had this bond to the dog, we had a relationship. We would take long walks in the woods and be friends, with the dog as a catalyst." Nancy now lives with her fourth German shorthaired pointer, whose name is Tomato, and the dog remains her primary link to her father. "Even after all these years, when we speak it's about dogs. He absolutely comes to life over dogs. It's how he connects to society."

Kathleen, a divorced mother, says that her six-year-old Australian shepherd, Oz, "is the bond between my son and me." He's their common ground: when Ian, who's now twenty, comes home from college, Oz is the first thing they start to talk about, branching out from him to other topics. He has a soothing effect on both mother and son: "It seems easier to talk about difficult things when he's around." He defuses strain, makes room for affection. "Ian was raised around a lot of love," Kathleen says, "and a lot of that was because of Oz. He just brought it out of us. He brought us out of ourselves." Oz is enormously fond of pancakes, and when Ian is home from school, Kathleen will make a big pile of them, give Oz his own stack on a separate plate. Simple story, possibly horrifying from a dental point of view (Oz's fondness is actually for maple syrup), but it gets at that connective effect: the mom, the son, and the pancake-eating dog in the kitchen is somehow a warmer, lighter image than the mom and the son alone. They laugh at Oz. They are co-conspirators in his indulgence. He gives them a mutual source of joy.

People often say things to a dog — or in front of a dog — that they simply can't say to other humans, particularly family members. Dogs can be objects of deflection, the animal you yell at, as my father did, instead of yelling at your spouse or your kids. They make wonderful objects of indirect communication, absorbing comments that one family member won't or can't say aloud to one another. ("Don't listen to him, he's just being a pig," a woman might say to the dog, the piggish husband within earshot but not addressed directly.) And sometimes, dogs are just the available pair of ears, the inadvertent object of free association. My friend Meg recalls a weekend from her youth, when a group of her parents' friends descended on the family home, near New Haven, Connecticut. The friends, all Yale alumni, went off to a big football game, and Meg's father, a lonely and rather isolated man, not a Yale grad himself, apparently felt alienated from the group. As his friends all headed out for the day, he turned to the family's black Lab and said, rather mournfully, "Well, Tucker, I guess it's just you and me." Meg has had that memory lodged in her head for thirty years, it so indirectly but so vividly expressed her father's loneliness.

Dogs, of course, often make better family members than humans: they can be far less judgmental, far less moody, far more faithful in their attachments, and they don't criticize your cooking. Anita, a forty-one-year-old librarian from Montana, has a cocker spaniel–poodle mix named Sparky. "Mom," Anita's daughter says, "sometimes

I think that dog is your best friend." Anita shakes her head: "Honey, some days that dog is my *only* friend." The dog is Anita's unwavering ally, the one being in her household who's consistently and genuinely glad to see her when she comes through the door at the end of the day, who always responds to her, who's always there. "She never complains or demands," Anita says, "and all she wants is to love me. It's so freeing to be with someone who makes no judgments, who thinks you're wonderful even if you are a mess. You know those days when you just feel like you are the biggest mess? When no aspect of your life is going right and you think no one could possibly ever love you? But then there is someone who does: the dog."

The dog, moreover, loves you in a particular way, with a kind of focus and constancy that's rare, if not unparalleled, among even the most devoted family members. Anita describes this when she talks about the difference between loving the dog and loving her children, now ages nineteen, twenty, and twenty-three. "The largest difference," she says, "is the way they feel about me. I know that my kids love me, and we have always had a very close and open relationship, but I also know that I love them more than they love me. I don't say that to be whining or anything — I think they love me as much as children can love a parent — but it's just a fact of life that mothers love their children more than they are loved in return. I don't expect to get that love back in full measure; I expect them to pass it on to their own families. So I know that my relationship with my kids is somewhat lopsided, and it is the nature of that relationship that they will leave home and form closer bonds with others as they develop their own lives. But with Sparky, this is the home that she has come to for the rest of her life. And if our relationship is lopsided, it is lopsided in the other direction. Her whole life is lived for the purpose of spending time with her family. When I leave, she just lives for the moment when I will return. I find it so comforting to have someone love me so much."

In a study of 122 families with dogs, Medical College of Virginia professors Sandra and Randolph Barker found that close to a third of the participants felt closer to the dog than to anyone else in the family; Sandra Barker, an associate professor of psychiatry, says she's not surprised at the depth of the attachment (she herself has four Lhasa apsos, who sleep at her feet while she works), but she was jarred by the number who ranked the dog as emotionally closer than human family members. Her understanding: "The dog has qualities that are sadly hard to find in human relationships," she says. "What other relationship in your life do you have where there's total acceptance, no strings attached, no I'll-love-you *if* — if you clean your room, buy me a diamond ring, take out the trash?"

A building contractor I know, owner of a big, lumbering black Lab who accompanies him on jobs every day, would find Barker's findings a little less surprising. "You know," he says, "I love my wife and I love my kids, but I *love* my dog." He's echoing Anita's sentiment: the dog provides a constancy of affection that family members simply don't; dogs are easier to feel close to.

And yet for all these same reasons — because they're just *there*, so present and available and uncomplaining — dogs can also be lightning rods for trouble, as easy to act out on as they are to love. Trainers see this happen all the time: they see the youngest kid in a family — lowest on the family totem pole — who gets picked on by siblings and responds by turning around and picking on the dog; they see parents acting out disagreements over discipline with the dog, mothers who want to coddle and indulge

240

the dog, fathers who want to beat him across the snout with a newspaper. In the dog himself they invariably see little mirrors of family style — walk into a loud, disorganized household where the TV's blaring and the kids are running wild, and chances are you'll find an undisciplined and unsettled dog, a four-legged reflection of familial chaos.

On occasion, trainers also see signs of deeper trouble: dogs in complicated family triangles, dogs in the dramas of dysfunction. One trainer told me about a woman who came to her because the dog was causing "marital problems." Seems every time she and her husband started to make love, the dog would start barking and growling, tearing at the bedspread. The husband — distracted and annoyed — would get up and banish him from the room. This created a bigger scene: the dog would howl and scratch at the door, and the wife, plagued with guilt, would get up and go comfort him. The husband would get angrier, the wife would get upset with the husband, the dog would end up back in the bedroom. Then the husband and wife would go back to bed without speaking to one another. What, the wife asked the trainer, should she do? The woman seemed far more worried about the dog's feelings of rejection than the husband's, and far more focused on keeping the peace with the dog than within the marriage. So the trainer gave her some practical suggestions — make it clear to the dog that he's not allowed in the bedroom; don't give in to him when he starts howling; once he figures out that he's not going to get his way, he'll shut up and adapt — and she kept her real assessment to herself. Inside she was thinking: Sorry, you don't need a dog trainer, you need a sex therapist.

Another trainer tells of a woman with an aggressive chow chow, a young female who was highly bonded to the woman but appeared to loathe every other member of the family: she growled and snapped at the woman's husband, wouldn't let her five-year-old son near her. The woman called in the trainer at her husband's request — he vehemently disliked the dog, worried that she might bite their son, said they'd have to get rid of her unless his wife dealt with the problem. The trainer sided with the husband: you've got an aggressive dog who dislikes children and men, she said, the behavior appears to be entrenched, and the breed can be stubborn and difficult to train; you're asking for trouble. Her advice: the woman should either put the dog down or find her another home, preferably with a single female who's not likely to have a lot of men around. The owner dismissed both suggestions out of hand, not simply because she didn't want to give up the dog but because, she admitted, she liked being the dog's "favorite" and found something gratifying about the way the dog elevated her in the family hierarchy. In the trainer's view, the dog not only gave the woman a sense of alliance she didn't get from her husband, she'd also become part of a family triangle, pitting spouse against spouse in a subtle drama about power and value. She needed that relationship with the dog, so much that she was willing to jeopardize her child's safety to protect it.

If dogs can help some people act out conflicts, they can help others mask them, providing ways to avoid looking at, or even feeling, problems or strains in a marriage. A paper presented at an American Psychological Association convention in Toronto, described a couple who connected profoundly, but solely, over their dog: they doted on the dog, talked about the dog, shared in the dog's care, and yet the dog also allowed them to maintain a degree of distance from one another: for example, their lovemaking was inhibited by the presence of the dog, who slept on their bed. When the dog died suddenly, so did their strategy: the distance and emptiness between the two, which they'd dealt with so successfully by focusing on the animal, rose to the foreground, and the couple separated shortly thereafter.

Dogs and human conflict: amazing how an animal can trot right in and sniff out trouble. Me, one of my central, defining struggles in relationships has always been a deep and abiding uncertainty about the question of what's *enough*. Did my parents love me enough? Do others? Am I good enough? Lovable enough? Can I get enough? The focus of this question — the object — has shifted innumerable times, but it's been with me always, a persistent free-floating anxiety about being disappointed and undernourished that keeps attaching itself to something new. For a long time it attached itself to food: in my twenties I went through an acute period of anorexia, starving myself down from 120 pounds to 100 pounds, then 95, then 83. In some respects I'd resolved the question — what's enough? — by bypassing it altogether, determining that I needed nothing, could get by with nothing, simply did not require nourishment the way others did. It was a beautiful solution in its own twisted way — if you have no needs, they can't go unmet — but it emaciated me in every respect, made me thin and sad and unbearably lonely, so in my early thirties I shifted the focus. I attached myself to Julian, a man who wouldn't love me, and thereby traded in physical anorexia for a more emotional brand. I saw this at the time as a central challenge, rather than a central struggle: if I could get this cold, critical, witholding man to love me, then I'd prove beyond all measure of doubt that I was, in fact, worth loving. If I could get him to feed me, I'd be safe, the battle won. In fact, Julian fed me table scraps — a hug here, a compliment there, a periodic crumb of validation — and it took me a long time, years, to understand that that wasn't enough, that living with him merely replicated, in exaggerated form, the household I grew up in. I started drinking heavily, another attempt to satisfy the same nameless hunger, and there was never enough of that, either. How to feed the self? How to get fed? What's the right amount of food or drink or contact or attention? I seem to have come into the world without the right internal gauge, some central mechanism that tells me how much is too much and how much is too little, so I've spent the vast portion of my life vacillating between deprivation and excess, yearning and claustrophobia, black and white.

At the time I got Lucille, this was the central struggle with Michael: How much distance to maintain, how much closeness? What was the right amount of either? If Michael strayed too far from my orbit, I'd panic with need for him — can't live without him, can't get fed without him, can't function. And yet if he got too close, assumed a certain level of commitment or permanence in my life, I'd be flooded with dread — he's too much, he's overwhelming, he'll consume me. What was missing in this equation, of course, was some central seed of faith, an understanding that the right balance between self-care and caring from others was achievable, a certainty that my needs could be met.

Lucille, I suppose, landed directly into the center of all that confusion: she became an emblem of my wish for an emotional sure thing, a relationship that would, in fact, be enough, a connection with another being that would be so vital and exclusive, no one else could have access to it. This is why I was so profoundly possessive of her, so jealous of her affection. She was the one creature on the planet I loved without reservation, and I had to have *all* of her love in return, for the alternative, I believed, was to be left with nothing.

So we'd sit in the couples therapist's office. "But she *is* yours," Michael would say. "The dog adores you."

I'd shake my head and try to hold back tears. I could believe this intellectually, but not at my core. I simply didn't trust love — hers, anyone's — enough to share it.

We saw the couples therapist together about half a dozen times before separating, and for a long time I wasn't entirely clear about what had happened: Did the dog merely highlight preexisting conflicts, underscore differences in our respective needs and wants and degrees of commitment? Or did I do something truly insane? Did I up and leave my boyfriend for a dog?

8

SURROGATE DOG

About six months after she got her puppy, an irrepressible malamute named Oakley, my friend Grace had lunch with a friend on Newbury Street, a chic stretch of restaurants and shops in Boston's Back Bay. Grace, who's about as in love with her dog as a person can be, brought along some pictures of Oakley, and when she pulled them out of her purse to show them, her friend pulled back from the table ever so slightly. "Oh, Grace," she said, raising an eyebrow. "Retreating into the world of furry animals."

Grace told me this story while we were walking in a stretch of conservation land about twenty minutes west of Boston. We'd stopped for a moment at the bank of a pond, and the dogs were engaged in a furious chase along the sand, darting up one side and tearing back, about as happy as dogs get. The sky was steely gray, the air and water calm, and we'd been out walking for about an hour, talking about dogs and the choices in life they help clarify. I remember that Grace stood there with her back to the reservoir and swept one arm out across the

251

scenery. "Retreating?" she asked. "*This* is retreating?"

Grace and I are both the kind of people that others have in mind when they talk about the tendency among some humans to use dogs as surrogates, to "retreat" into the world of animals in order to bypass more problematic and complex human relationships. I can see the thinking behind this view: Grace and I are both single women who live alone, work out of our homes, invest extraordinary amounts of time and energy in our dogs. We are both prone to periods of isolation and withdrawal, people who might very well prefer lounging around at home with the dog to hanging out at some swank café on Newbury Street. Neither of us has kids. And me, I've left this seven-year relationship with a genuinely good man to spend most evenings holed up in my living room, the dog at my feet. So it's not hard to see why we'd be looked at a bit warily: at least superficially, we appear to have made a rather deliberate set of choices — dogs instead of people, dogs instead of children, dogs instead of men.

And yet there we were, two women intimately engaged in conversation, sharing time and the natural world and our mutual love of animals. Grace, whom I met right around the time I separated from Michael, is a woman of uncommon intelligence and depth, and I doubt our paths even would have crossed had it not been for the dogs: we share the same dog trainer, went to the same dog camp in Vermont, fell into the whole dog world at roughly the same time. In the two years since we met, we've racked up countless hours together in those same woods, and our walks together have become one of the most sustaining aspects of my life, weekly shots to the soul of connection and laughter. We are very much on the same road out there, both of us going it alone in the world, trying to chart courses for ourselves that feel meaningful and true, and aware of the extent to which both our dogs and our friendship with each other have factored into that effort. Like she said, *This* is a retreat?

As a culture, we're a bit schizophrenic when it comes to loving dogs, both accepting and suspect. On the positive side, keeping a pet — particularly a dog — can grant you a stamp of normalcy, give you a casual but handy entree into the social world. A well-known study by New York psychologist Randall Lockwood suggests, almost without exception, that people in the company of a dog are more likely to be re-garded by others as friendlier, happier, more relaxed, and less threatening than people who are dogless; in an oft-cited study on the social effects of keeping a dog, British zoologist Peter Messent found that dog walkers in public parks and gardens had higher numbers of positive interactions and more extensive conversations with others than people who were either on their own or with small children.

And yet some fine line exists between "normal" love for a dog and "excessive love." Care for the dog too much — dote on the dog, spend too much time with the dog, get too attached to the dog — and you get branded as something very different: you're eccentric, or antisocial; you get laughed at. In his book, *In the Company of Animals*, James Serpell traces part of what he sees as the cultural denigration of pet keeping to the popular press, which seems to devote as much space to pet-human relationships as it does to people's sex lives, with the bulk of the coverage designed to highlight the ex-tremes to which today's pet owner goes. There are stories about pet cemeteries and pet summer camps; stories about the modern accessorized dog, with his gold choke chain and Burberry raincoat and special-ordered, hydrant-shaped birthday cake; stories about excess. A classic example, the kind of story the media love to laugh at: Countess Carlotta Liebenstein, an eccentric German noblewoman, who left an estate valued at $80 million to a German shepherd dog called Gunther. The overt message here is clear — people who love animals are wacky — but behind it is a more covert and subtle one, a belief, as Serpell describes it, "that pets are no more than substitutes for so-called 'normal' human relationships."

In fact, very little evidence exists to suggest that people with deep attachments to their animals are any "weirder" than people who are less attached, or that they're focusing an unhealthy degree of social energy on their pets. If anything, dogs tend to widen, rather than narrow, one's social world. "Morgen finds it easier than I ever did to go up to strangers and introduce himself," says Bill, of his three-year-old dachshund. Single and in his fifties, Bill has lived in a high-rise condo in Washington, D.C., for more than ten years. Pre-dog, he hardly knew any of his neighbors; now he knows dozens of them, Morgen having wagged his way into a vastly expanded circle. Pre-dog, Bill led an active social life; today it's doubled: he chairs his condo association's pet committee; his friends' children, who love Morgen, make regular "play dates" with him and the dog; several of the acquaintanceships he's made through Morgen have blossomed into close friendships, particularly with other dog owners.

This is a classic story: dog gets owner out of house, tugs him or her toward new people, expands the human pack. Lisa, a school administrator who owns a small black dachshund mix named Franny, first met Mimi, a social worker who owns a small black miniature poodle named Marty, through a dog group; their dogs became playmates, the women became intimates; when Mimi became pregnant eighteen months later, Lisa became her labor coach. Jonathan, who owns the basenji named Toby, met his current lover, a veterinarian named Mike, through the dog. Dog love became human love, colliding at key points (on their third date Mike turned to him and said, "Jonathan, I really love being with you, but I've gotta ask you to stop bringing me Toby's stool samples."). Like Bill's dachshund, Jonathan's dog gave him a sense of belonging in the world. "I walk with this whole network of people who own dogs," he says, "and I've found this really wonderful community. We all walk our dogs in the morning and the evening, and sometimes I see them in other places, and we call each other by our dogs' names: 'Oh, there's Toby's father, there's Astro's father.' "

Such stories confirm what researchers have documented many times over: dogs are excellent social lubricants, and they tend to attract relatively social people. Psychologists at the University of Oklahoma have found that people with affectionate attitudes toward their dogs have proportionately affectionate attitudes toward people; British researchers have reported that people who interact frequently with their dogs have a higher desire for affiliation with other people than non–dog owners; a California study reported that elderly pet owners were more self-sufficient, dependable, helpful, optimistic, and socially confident than non–pet owners.

I suppose the key in human-dog relationships — at least as others see them — is degree. Certainly the world feels like a more comfortable, social place to me when I have Lucille by my side: passersby smile, sometimes stopping to ask a dog question or two; I tend to feel more relaxed and less anonymous when she's with me, and also more approachable, my four-legged ice-breaker at the end of the leash. And yet I'm also aware of that fine line, that question of excess, that view of what's "normal" and what's not. The dog clearly does not occupy a secondary or neatly comparmentalized role in my life, so little seeds of doubt periodically crop up inside: I seem to love this dog too much; is this a problem?

On Christmas Day two years ago, I showed up at my aunt's house for the afternoon, Lucille in tow along with a big bag of Lucille's stuff: a blanket for her to lie on, a couple of chew toys, a big old rawhide bone for her to gnaw while we ate Christmas dinner. I felt a little silly lugging in all that gear — it felt like the canine version of a diaper bag — so I kind of tucked the bag under my arm, then walked into the living room and looked around.

Christmas is a lonely time for me, particularly since my parents' deaths. What is family? Who in the world do I really feel connected to? All those dark existential holiday questions bubble up, and they do so with particular intensity in the aftermath of those losses. I've spent every Christmas since childhood with my aunt and her family, but it's not a group I see much of between holidays, and I remember standing at the entrance to the living room feeling orphaned in the truest sense, as though I were about to spend Christmas with a large group of people who didn't really know me very well. So I hovered for a minute, and then homed in on my cousin Suzanne and her husband Bill, who'd gotten a puppy right around the time I got Lucille, a standard poodle named Pepper.

Oh, good, I thought. Common ground. We said hello and exchanged brief pleasantries, and then I asked Bill, "So how's Pepper?"

Dogs are one of the few subjects I can get truly gabby about, so I think I hoped we'd launch into dog talk from there, trade stories about training and behavior problems and care and feeding. But Bill paused just slightly and said, "Oh . . . um, she's fine. She's turned into a really sweet dog." Then he gave me a kind of blank look, as if to say, "Next question?" and I remember being struck by a sense of acute awkwardness, standing there with my dog and my dog gear and my dog question. Here I am: dog, dog, dog.

Suzanne and Bill, both physicians, have two young daughters, and they lead very busy lives, and although I'm sure they're very attached to Pepper and are pleased that they got her, she does not occupy a primary role in their world. By contrast, I'd spent the whole day absorbed in my dog — we'd gone for a three-hour hike in the woods that morning, and I'd carted her along with me to Christmas dinner as though she were my daughter or my date, and I felt hugely exposed for a second, as though I was revealing some fundamental difference between me and other people: woman with dog versus man with family. Woman with tiny narrow life versus man with big full life. Woman with bizarre priorities versus man with normal priorities. What's wrong with this picture?

This brand of self-consciousness can hit people who love their dogs deeply, even when they're together with like-minded dog devotees. I was hanging out at a park recently with a woman named Catherine, who owns a yellow Lab named Bailey, and a teenage girl named Katie, who owns a golden retriever–Lab mix named Sadie. At one point Catherine pulled out a little container of treats from her knapsack, then turned to the dogs, who were milling around the picnic table where we sat, and said, "Okay! Who wants a snicky-snack? Would anyone like a nice snicky-snack?" Her voice was squeaky and high, as though she were addressing a flock of schoolchildren, and hearing herself, she looked at Katie and me and rolled her eyes. "Oh my God," she said. "What is wrong with me?" We just laughed: happens all the time.

And it does happen all the time. Like a lot of dog lovers, I have about fifty different terms of endearment for Lucille — sweet pea, and Miss Pea, and pea pod, and peanut, and Miss Peanut — and every once in a while I'll hear myself in the house cooing at her in an overenthused soprano, "Oh, hello, you sweet, sweet pea! Are you the sweetest pea there ever was?" and I'll just pray my neighbors can't overhear me, I sound like such a goon. Or I'll be making her dinner, and I'll catch myself imploring her to eat as though she's a toddler — "I have a *delicious* supper for you, Miss Pea! Have a bite of this delicious supper!" — and I'll shake my head: Good Lord, I have gone off the deep end at last. I'm alone with the dog, and I'm alone with her a lot, and so the question of substitution looms large and often: What am I doing here? *Is* she a surrogate for other relationships? *Should* I be investing all this energy elsewhere?

Should: that's the key question, the same one that generated the sense of exposure I felt at Christmas. Should I be living this kind of life or some other kind of life?

Should I follow a more traditional path, pursue more traditional goals, let go of that leash and follow not the dog but, like my cousin and her husband, a life that includes marriage, kids, a home with actual people in it? And if I don't follow that path, does that mean there's something wrong with me?

I have been pondering these questions almost since the day I got Lucille, and I'll no doubt continue to wrestle with them over time. On bad days, days where I'm lonely and my world feels small and unproductive and gray, I lean toward the pathologizing view, look around and see myself as some sort of reclusive, dog-obsessed misfit, too fearful and damaged to lead a "real" life. But other times I'm less sure of that. Alongside the seeds of doubt I felt cropping up at Christmas, there was also a small seed of certainty: this dog is an enormous solace to me, a constant companion and witness to my daily life, a being I have come to feel closer to in many ways than members of my own family. She represents a choice, a style of living and loving that may not be conventional but that is valid in its own right, if only because it's my own.

A month or so before we began to separate, and several months before we went into couples therapy, Michael and I spent a week together in Vermont, where we had one of the worst fights we've ever had. We were walking with Lucille through a piece of conservation land in the Green Mountains, and we'd stopped at a hillside that over-looked a wide, lush vista: trees, farmland, a sparkling jewel of a lake in the distance. For reasons I can't recall, I was in a snippy mood, snippy and mean-spirited, and I started to babble about this question of surrogacy, about how annoying I found the implication that dogs are somehow low-rent versions of children, poor substitutes for people who aren't noble enough or brave enough or somehow normal enough to have kids.

And then I blurted it out: "The fact of the matter is," I said, "I don't want kids."

Michael and I had talked about children over the years in the same noncommittal way we'd talked about marriage: a kind of maybe-someday way that didn't open the door very wide but never entirely shut it, either. In other words, I'd never said any-thing that declarative about children, and I was aware that my comment was barbed and thoughtless even as I said it. *There: don't want 'em; case closed.*

Michael didn't respond at the time — he

just sort of shot me a look — but later that day something triggered his anger, and he exploded at me. We were inside by then, in the house we'd rented for the week, and he brought up that comment — so simple and flip on my part — and he was (justifiably) furious: there I was, in his view, making this major statement about the future all on my own, with no concern for his desires; there I was, shutting him out; there I was, telling him in essence, "It's me and the dog, not me and you, and certainly not me and you and our future children." Michael, who would very much like to have a wife and a family, stood up at one point and said, "Listen, let's just finish up the week and then go back to Boston, and then you go your way and I'll go mine." Michael had never said anything quite that declarative, either, and I remember that I stood there and felt partly relieved but mostly terrified.

An hour or so later, the two of us having railed at each other for a good long time, I went off for a walk by myself and wound up back at the same spot where I'd made the initial comment. I sat on a rock, Lucille by my side, and I looked out over the farmland and the lake, and I panicked, panicked with the singular brand of dread and uncertainty that hits when you see your life at a crossroads, one path leading in a familiar direction and one path leading someplace entirely new and uncharted. I suppose it's safe to say that I'd already contemplated leaving the relationship for some time, that my difficulty in sharing Lucille with Michael had already spoken volumes about my view of us as a pack, but I'd always framed the issue in somewhat narrow and specific terms, the focus placed rather squarely on him: Do I want to be with *him,* could I have children with *him,* could I live with *him?* What are *his* strengths and limitations? That afternoon, facing what felt like the real possibility of life without Michael for the first time, I suppose I had to broaden that focus, to open it up, to look not just at who Michael was and what he might or might not be able to provide but at who *I* was, and what kind of a life *I* might or might not want to live.

I watched Lucille, who was ambling along the rocks, poking her nose into this crevice and that one, and I thought about kids. Like most women, I grew up simply assuming I'd have them, taking the matter pretty much for granted: you're female, so at some point the right guy sails in and you settle down and you have children. In fact, I've never had so much as a tingle of maternal desire, at least not toward children. When we were in our twenties, my sister used to talk about how her whole body seemed to be screaming to have a baby, how every cell in her was agitating for it, telling her it's time, do it, get pregnant, *now*. Me, I've always melted at the sight of a puppy, but I can't ever remember having that feeling toward a baby, not even once. For a long time I worried about this — was I missing some key procreative gene? did I pickle away my maternal instincts during all those years of drinking? — but I also figured the desire would come in time, get washed up on the same elusive tide that washed up the right man, that I'd want a baby as soon as I found someone I wanted to have a baby *with*.

Sitting there on that rock, I realized that I might never experience that kind of gut-level longing for a baby, that it just might not be part of my makeup, that having a child — or even wanting one — would require a shift in orientation so massive and fundamental it would necessitate something like a wholesale personality change.

I love taking care of the dog. There is no question that she satisfies some brand of maternal feeling in me, that I take a great deal of pleasure in being able to provide for and nurture her. But even when I'm carting around bags of toys and blankets for her, even when I'm down on my knees cooing at her or imploring her to eat her supper, there is no question in my mind that she is a dog and not a baby, that the kind of care I give her is very different from the kind of care one gives an infant or a toddler, that the sacrifices and rewards I experience are wholly different from the sacrifices and rewards of motherhood. Janet, the ER nurse who owns the yippy Pomeranian–terrier mix named Kim, stated the difference very simply: "You can't put a baby in a crate, toss in a couple of Milk-Bones, and go shopping for four hours. This distinction has never been lost on me." Indeed. The dog owners I know may put a lot of time into dog care, but we certainly don't log the kinds of hours that mothers of children do: no middle-of-the-night feedings, no crying jags on the dog's part, no temper tantrums, and — our own devotion and intensity aside — nowhere near the range or volume of worries. "I love kids," says Barbara, single and thirty-five, owner of a cocker spaniel named Joelle. "But let's face it. Your dog isn't going to turn to you as soon as he hits adolescence and scream at you that he hates your guts." Right again: dogs may represent the one relationship in a human's life where the potential for chaos actually diminishes over time. You don't have to worry about whether the dog will get into a good pre-school, or how you'll pay for his college education; you don't have to imagine the dog showing up one day with a mohawk or a substance abuse problem, and you can live in certainty that the dog will never crash the family station wagon into a tree.

I thought about that from my perch in the Green Mountains, too. I thought: If I worry this much about a dog — if I'm this caught up with her, this prone to projection, this concerned with her health and safety and emotional well-being — just imagine how I'd be with children; I'd have to keep them in a plastic bubble until they turned thirty-five. I'm not sure I have the capacity for distance, for loving detachment, for whatever it takes to let go of the parental leash. I'm also not sure I have the patience, or the stamina, or the enormous selflessness that parenthood requires, and I'm not sure I could tolerate the trade-offs and the million daily compromises it involves. "You feel different when it's your own baby." People always say that, that the requisite degrees of commitment and love and patience and selflessness just well up in you as soon as you're staring into the eyes of your own child (my own mother, who experienced the same lack of maternal flicker when she was in her twenties, used to say it), but I'm not so sure I want to find that out, not so sure I want whatever levels of commitment and love and selflessness I have within me to be directed toward a baby in the first place.

Teasing out the voices: that phrase came to me as I sat there. Here I am trying to tease out these voices in my head. The self-deprecating voice is loud, sometimes merciless: I'm somehow not good enough to be a mom, it says; I'm not equipped for it, not sufficiently selfless or well adjusted. This is the voice of our culture, the voice of our parents, the voice of 25 million child-rearing American couples, and its words are very clear: noble, normal folks have kids; selfish, screwed-up ones don't. Against the roar of that message, I think I've always heard another voice, tiny and tentative as a whisper, a mere thread of feeling inside that's said: But wait, I'm not sure I want that, not sure that's really the right path for me, not sure that's where my energy and talents would be best directed. Something about the dog, something about having her and loving her and caring for her, has given life to that whisper, made it clearer and more audible. I looked at her and thought: There are many ways to be nurturant in this life, many ways to be generative and loving. I thought about a slightly catty bumper sticker I once saw and secretly admired: CHILDREN ARE FOR PEOPLE WHO CAN'T HAVE DOGS. I thought: Perhaps, for me, this is enough.

Enough: that old, familiar question about what's enough. I called the dog over to me, and scratched her on the chest, and thought about our daily rituals: the way she pads up the stairs beside me every morning when I head into my office to work; the way she bolts to attention when I turn the computer off; the little games we play — a set of high-tens, a little chase around the living room — before we set off on our afternoon outing. I thought about the exquisitely simple joy of meeting her exquisitely simple wants: handing her a rawhide stick, which to her is the equivalent of a drug; presenting her with a new toy; providing her with exercise and sustenance and affection and companionship. And I thought, with a combination of relief and great surprise: Well, yes, in fact, for now this may be enough; at last, this feels like enough.

One of the great surprise bonuses of my entry into the world of dogs has been the discovery that plenty of other women have had the same experience: in the great should-I-or-shouldn't-I conundrum around children, dogs can provide a missing piece of the internal puzzle. Sometimes a dog will tug a woman toward the yes-I-should side of the equation. ("I didn't know I had it in me, this ability to bond and love, until I got the dog," says Amy, a thirty-seven-year-old editor, whose earlier indifference about children has given way to a more concrete longing for them.) Sometimes a dog will have the opposite effect. (From Sandy, forty, a lawyer, self-described control freak, and owner of two standard poodles: "Infancy and breast-feeding and up all night? No thanks. I barely made it through puppyhood.") And sometimes a dog will fill the gaps — and quite ably — while a woman waits for time and circumstance and her own heart to pull her one way or another. ("I'd love to have kids someday," says Kathy, twenty-nine, owner of a springer spaniel named Milo. "But for now I'm perfectly happy being a dog mom.")

Like me, all of these women are single and devoted to their dogs. Like me, none of them are confusing their dogs with human children. And like me, all of them resent the implication that there's an element of sublimation at work, that the time and emotional energy they spend on their dogs represents a deflected wish for children. Amy finds the idea ridiculous: "I love kids and I love dogs, but come on: they're completely separate." Sandy finds it sexist: "Do men with dogs get accused of this? Can you imagine seeing a guy walk across a field with his German shepherd and having your first association be, 'Oh, he's obviously sublimating his biological need to procreate into that dog'?" Sandy laughs out loud at the idea, then grows more sober. "What I really resent," she says, "is the idea that there is only one way for a woman to make a contribution in her lifetime, by having kids." She herself is a paragon of giving: works extremely hard for her clients, does regular pro bono work, supports half a dozen social causes, maintains an equal number of close friendships, *and,* as she puts it, loves her dogs to death: "So where's the sublimation?" she asks. "These dogs aren't my kids. They're my friends. They're my fun. I love taking care of them. I love how happy they are when I walk in the door. I love having them in my room with me when I go to sleep at night. I love watching them *eat.* Do I sound like a mother? No, I sound like a woman who loves her dogs."

Well, truth be told, she sounds like both:

a woman who loves her dogs in a doting, smitten, maternal way, the same way I love Lucille. A lot of dog owners I know employ that style — verbal and cooing and affectionate — but does that make our dogs a substitute for human children? If motherhood is seen as a biological imperative or a symbol of normalcy or a necessary rite of passage, then perhaps, I suppose so. But if it's seen as a choice, stripped of judgment, then no: the dog is simply — and blessedly — the dog.

And yet man cannot live by dog alone. The panic I felt as I sat there on that rock in the Green Mountains had less to do with the possibility of not having children than it did with the possibility of not having a partner, of leaving the man with whom I'd shared seven very defining years, of choosing to be alone.

This is something I've never really done before, chosen to be alone. I've been alone because men have dumped me, and I've been alone because I've dumped men, and I've been alone (a lot) with my various addictions, holed up at home over the years in order to drink in secret or to starve in secret or simply to avoid being out in the world with other humans. But all those stretches of solitude have felt impermanent and involuntary, marked by great strains of longing, as though I was merely biding time, waiting for my life to begin, waiting for some great elusive external to come along — new job, new man, sudden personality change — and alter everything, give my life meaning miraculously and immediately, just like that. There is enormous passivity to that brand of solitude, that kind of waiting and waiting and waiting, and there is a profound form of misplaced hope behind it, a belief that change and happiness and solace are things that simply *happen* to you, that they come from the outside in, little flukes of luck or circumstance or twists of fate that simply descend from the heavens with no agency or determination on your part, nothing deliberate about them.

So I contemplated deliberation, choices. Life without Michael. Life without a partner. Self-definition without a man and without parents and without white wine, glasses and glasses of it firmly in hand. I sat there and I thought: *What would I do?* I meant that in the most specific and literal sense: Without Michael, without a boyfriend around whom to plan evenings and weekends and vacations, how would I spend my time? How would I fill the hours? Left to my own devices — left with my own barely understood needs and fears and longings — what foods would I eat, and what activities would I find to occupy myself, and who would I spend time with, and how lonely would I feel, and how bad could things get, and who in the midst of all that — just *who*, in the end — would I turn out to be? A strong person or a fragile one? Passive or powerful, social or solitary, capable of caring for myself or inept, balanced or crazy or *what?* I sat there in the mountains, and I thought: I'm thirty-six years old, and I still have no clear answers to those questions. I thought: It's time to start. And then I thought: The dog, at least I have the dog, thank God for the dog.

Lucille as companion. Lucille as agent of structure, the being who'd get me out of the house each morning and night, out to the woods each weekend. And Lucille as a kind of guide dog, the creature who'd be by my side as I began to create another kind of life for myself, to look at what it might mean to create a life in the first place.

Michael and I returned from Vermont at the end of July, then decided to spend August apart, a kind of trial separation. August led to September; September led to couples therapy; couples therapy led, by

Christmas, to a more final decision to part. Did I leave him for the dog, or because of the dog? Not really. Was I able to leave him because her presence, and the joy and comfort she provides, made me feel safe enough to move on? More likely.

But questions linger, primarily about isolation. The dog may have helped me get out of a relationship, but could she ever help me get *in* one? Or have I found in her some kind of alternative to adult intimacy, a less complex and demanding way of living with another being?

A few months after Michael and I split up, I had a long conversation about this with a woman in San Francisco named Ros, a journalist who lives with a five-year-old dalmatian–Lab mix named Digby. Digby is Ros's Lucille — beloved animal, primary relationship, center of her universe — and like me, Ros organizes a good chunk of her life around the dog: hates leaving the dog alone, spends most of her free time with the dog, has cut way back in the four years since she's had her on non–dog activities, like movies and dining out. So she wonders about the question of surrogacy, too. Ros, single and approaching forty when we talked, asked aloud: "Should I be concentrating on having better relationships with people? Is my relationship with the dog just sort of holding things up, helping me avoid venturing out into the human world?" For the most part, she thinks: No, the relationship with Digby feels like an enormously healthy and sustaining force in her life, probably the healthiest relationship she's got going, but the worry gnaws at her periodically, creeps up in moments of worry that I understand completely because I share them.

Sometimes, at the end of the day when I bolt that door behind me, take my deep breaths, and acknowledge just how much fear I lug around, I can see that my relationship with Lucille does have the quality of a retreat, a sanctuary where I'm temporarily eased of the burden of grown-up, human interactions. Sometimes, when I'm turning down an invitation to do something human (a dinner, a movie), I can feel it, can feel myself casting a vote for solitude, surrendering to my own need for safety. Easier to be with the dog. Just want to stay home with the dog. That's when the questions bubble up: If I spent less time with the dog, would I have better, richer, deeper relationships with people? Would I be (gasp!) dating? Am I shielding myself from some other kind of experience here? Where, in the end, is the line between self-protection and self-limitation?

I don't think there are clear or easy answers to those questions. Right around the time I spoke to Ros, I had coffee with a woman named Marjorie, sixty-five, a recently retired school administrator who has lived with border collies for the last thirty years. Marjorie has never married or had children, so she has spent a lot of time contemplating the question of surrogacy, wondering about that same line.

At the moment, Marjorie is sharing her life with a two-year-old border collie named Cory, who is sprightly and agile and, true to his breed, extremely intense, equipped with a blend of concentrated energy and focus that's startling if you're not used to it: throw Cory a ball, and he'll dash after it as though his very life depends on the retrieval, then careen back to you, drop the ball at your feet, stand there, and stare at you, front legs splayed, head tucked low, his whole body quivering in anticipation: throw the ball, throw the ball, do it again, do it again, *please do it again*. Marjorie is very attached to Cory, doting and adoring — "He's turning out to be an extremely satisfying dog," she says — but if you ask her to talk about what it's like to love a dog profoundly, she will tell you about her first dog, a border collie named Glen whom she acquired when she was thirty-three.

"Glen," she says. "From the day I got that dog, we were absolutely together. We were just a perfect match: it was like, I don't know — a foot in a shoe. I sometimes thought of him as my other half, or my alter ego." Glen apparently felt the same way: when Marjorie went to work, he'd periodically escape from the house and literally track her down, navigating his way across two towns and several heavily trafficked roadways before materializing outside her office; he couldn't bear to be separated from her. Marjorie says she hasn't had such an intense relationship since, with a person or with another dog. "I hesitate to say this," she said, "but I think in some ways Glen was like my husband — not in a sexual way, obviously, but emotionally. And I think the relationship was so intense that it satisfied me — it satisfied a great need to love something and to be loved — and to some extent I didn't need anyone else. I just plunged into my relationship with him. I didn't hold back at all."

Glen lived to be fifteen years old; he died when Marjorie was forty-eight, and was succeeded by four more border collies: first Kate, who was Glen's daughter; then a pair named Bobby and Jamie; and now Cory, her current dog. Marjorie is a lively, genial woman with a wonderfully robust laugh, the kind of person who projects great vim and extroversion but is actually quite private, shy, and self-conscious. When she's with a group of people, particularly people she doesn't know well, she has a hard time turning off the voices of self-criticism, the harsh judgments: is she smart enough, is she adequate? *Oh, you jerk; you sound like such a jerk:* that's the kind of thing she hears in her head, a burden that's blessedly absent when she's at home with a dog. Social skills and social confidence, like muscles, atrophy when they're underused, so part of this is circumstantial: Marjorie has lived alone since she was twenty-nine, and she's simply fallen out of the habit of being in social situations, forgotten how to be comfortable with small talk and easy conversation. But she also worries that her reclusiveness runs deeper than that, that it speaks to some darker failure on her part.

"I'm not lonely," she says, "but sometimes I think: 'What the hell did I do wrong in my life? Why am I alone? Why don't I have a family and children?' I always thought I was going to have all that, and I can get pretty down on myself, thinking: 'Oh, you can do fine with dogs but you can't have relationships with people.' "

It is so hard to assess motive, to define what really drives a person to make the choices they make. Marjorie sometimes thinks that the dogs caused her to become more isolated than she would have been otherwise, that if she hadn't had Glen, she might have been more motivated to propel herself out into the world, to look for solace and companionship among people. And yet she's not at all sure that would have happened. Her career, dominated by women, didn't bring her into contact with a lot of men. She was never a very gregarious, social person, not the kind of person who would have thrown herself with anything approaching ease into new social situations, into parties or restaurants or dating services. So the same forces that made life with Glen so satisfying — the need for solitude and self-protection, the wariness about the social world — might have been manifest in some other form, kept those same doors shut on their own accord. "If you're worried about being isolated," Marjorie asks, "do you go out and meet people or are you just isolated? In my case, it's quite possible that

I would have gone on being single and alone. And then I wouldn't have had either — not the family and not the enormous gratification of the dog."

There are so many ways to carve out a life for oneself, and yet how persistently we hold to the idea that there is really only one way, one valid path. I could hear strains of that feeling in Marjorie's words; there was an either-or-ness behind them, as though only two roads had stretched before her, a family road and a dog road, and she'd opted for the latter, chosen something fake instead of something real. And yet, perhaps unintentionally, her words also belied that logic.

Marjorie used the words "other half" and "partner" to describe her relationship with Glen, her first dog. When she described her next three dogs — Kate, Bobby, Jamie — the words "children" and "kids" came up. Bobby and Jamie were like two little kids, two kid brothers — different from Glen, her relationship with them less intense and more playful. And Cory, her current dog? "This dog feels like a grandchild," she said. "He's kind of a permanent baby." Marjorie didn't use those descriptions consciously — she didn't sit me down and say, "First I had a dog husband, and then I had dog children, and now I have a dog grandchild" — but as I listened to her, I was struck by the idea that she'd created her own sort of family structure through those dogs, that they'd fit into the arc of her life in many of the same ways family members do, and that there was a lovely sort of symmetry in that effort. Marjorie may have regrets, may worry about the trade-offs she's made and the experiences she's missed — who doesn't? — but she's also constructed her own path, a private road of powerful attachments and familial feeling and no small measure of solace. In her own way, Marjorie has found a solution. As she put it, "Thank God for the dogs."

I, of course, am not immune to the idea that there's only one way to live a "normal" life, so Marjorie both reassured and alarmed me. I sat there and I listened to her and I wondered if all these choices of mine — getting the dog, leaving Michael, coming to favor the world of woods and dog parks over the world of restaurants and parties and future children — represented the first steps down a path like Marjorie's, a private road with a dog at its center.

I am mixed about this possibility; I vacillate. I ask myself: Is the dog keeping me from some broader exploration of what's out there? If I didn't have her, would I really throw myself into the social world with more vigor, go to Italy, take up needlepoint, enroll in a karate class? Or is she what makes my current exploration, internal and solitary though it may be, comfortable and sane? I worry: Is the dog a balm against isolation or an excuse for it? A symbol of what I fear, or a symbol of who I am? I hear myself getting judgmental, succumbing to self-criticism — am I crazy or am I sane? normal or abnormal? either-or — and when those voices get too loud, I try to view the matter through a different lens. Maybe — just maybe — the dog is something else entirely.

Grace and I often talk about the possibility that dogs don't dictate options so much as they help illuminate them: dog as agent of elucidation, dog as vehicle for self-definition. The first time we went for a walk together, Grace said in a perfectly matter-of-fact tone, "Of course, dogs are a metaphor for change," and I knew then and there that we'd become friends.

Grace, a painter, is forty-two years old, tall and striking, with auburn hair and high cheekbones. Like me — and like Marjorie — she's a bit of a recluse, the kind of person who'll stand in the middle of a social event looking perfectly comfortable and elegant but quietly sinking into despair. Outside versus inside. Grace is a very introspective woman, she loathes small talk, and although she can slip masterfully into a polished social persona (she has the right little black dresses and the right jewelry and shoes and all the right techniques required to make other people feel important), she has also racked up untold hours at gallery openings and dinners secretly thinking: What am I *doing here?* I have nothing to say to these people. Those thoughts lead to darker thoughts: There must be something wrong with me. Why is it that everyone in this crowded room appears to be having a good time while I'm standing here feeling alienated and estranged, wishing I were home watching *ER?* Grace lives and works alone, and like me she can pass many, many consecutive nights — five, eight, ten in a row — without a lot of social contact. This worries her, too. She wonders: Am I a hermit? A misfit? Do I have a life?

Grace lives in a scruffy blue-collar neighborhood of Boston, and these days she often finds herself tramping around the streets at night with Oakley, out for their evening walk. Oakley is a spectacular-looking animal — wolflike, with a black mask and a massive coat — and all manner of people stop to talk to her: big guys with tattoos, mothers with children, other dog walkers. She'll chat, she'll answer the predictable questions (no, she's not a wolf; yes, I have to brush her a lot), and then she'll wander off down the street, and every once in a while she'll stop and compare this version of herself out in the world, the dog-walking version, to the version she inhabited pre-Oakley. No cocktail party chitchat; the small talk is about something she actually cares about, the dog. No glasses of white wine or little canapes, either; she's holding on to a leash, and that seventy-four-pound animal at the other end of it gives her a sense of power and comfort and safety so palpable, it takes her breath away. The dog makes her feel connected to the world in an entirely new way, a way that feels easy and secure, a way that feels *true*. Grace will stand there and think: But I *do* have a life. Sometimes it's a solitary life and sometimes it's a lonely life and certainly it's an unconventional life, but this is it, a life with a home and a dog and work and friends, a life that finally feels right.

Feels right: music to my ears. My therapist has tried to steer me toward that feeling for eons: forget about what you think you're supposed to do, forget about what others expect you to do; what feels *right,* to *you?*

That's the hardest question, because like the question about children, it means separating out so many opposing voices, trying to pay heed to the one that lies hidden at the center.

Like Grace, I'll sometimes find myself walking through downtown Boston or Harvard Square, in Cambridge, and I'll see people out and about, people who look like they're leading real lives: families with children, or couples strolling hand in hand, or groups of friends huddled at the window table of a restaurant. I'll struggle for a moment with the voices of social normalcy: Should that be me, that woman with her baby carriage, or that one with her boyfriend, or that one eating out with her ten best friends; should that be what I'm aiming for? And at that core level, the one that's most fundamental and true, the answers are emerging, the internal voice growing more resonant: I am a person who's held babies and felt nothing, who's walked hand in hand with a man and felt utter disconnection, who's sat with ten friends in a restaurant and thought: I am so lonely. But the dog: the dog touches some other piece of me, some nascent place inside that feels not only more solid and real but also more open to connection.

The day I met Marjorie, I'd gotten up and walked around Fresh Pond with my friend Wendy, who's probably cruised that same two-mile loop with me a hundred times. Wendy — fifty-four, a lesbian, deeply involved in women's health care — is the sort of friend who'd have never crossed my path were it not for the dogs, the circles we travel are so different, but we met when our dogs were puppies and started walking together on weekday mornings, gabbing away while the dogs chase each other through the brush or amble along beside us. That afternoon I hooked up with my friend Hope, inhabitant of another circle distant from my own (she's an atmospheric chemist, Harvard affiliated), and we sat at the dog park as we do most afternoons, talking while the dogs played. In the evening I made a couple of phone calls — left a message for my friend Tom, a dog-owning writer friend, and asked if he'd like to walk with me on Saturday; spoke to Grace for a long time, made plans to go to the woods that Sunday.

This is a lot of dog time — hours walking the dog or sitting while the dogs play — but it's also a lot of human time. Wendy, who sees me before seven A.M., in the hours before my defenses have had time to gel, has witnessed me at my most vulnerable and tender: we have walked together on our birthdays, on the anniversary of our parents' deaths, on good days and bad days and mediocre days; we have walked through rain and snow and sunshine, and through the small victories and defeats of our daily lives. Hope, whom I meet at the end of the workday, probably knows more about the vicissitudes of my life as a writer than anyone: in the hours we've racked up together at the park, we've talked a great deal about dogs but we've also talked about work and self-esteem and the link between them, about depression and relationships and families, about the daily challenges of getting by. Tom and I got to know each other through our dog trainer. His father, like mine, died of a brain tumor the summer we met, and we spent long afternoons marching through the woods west of Boston with our dogs, talking about loss and the way it changes you. Death marches, we called them. And Grace — meeting Grace has been like discovering a long-lost sister, a kindred spirit who's been out in the world all this time forging a nearly identical path.

And then, of course, there is the dog herself; there is Lucille. That evening, after I'd locked up and fed the dog and made my phone calls, I climbed into my bathrobe and curled up in a chair in the living room, ready to settle down for the night with a crossword puzzle and the television. I pass a lot of evenings this way, hunkered down in that chair, and sometimes I'll look up from whatever I'm doing, and I'll watch Lucille for a moment or two. She always knows when I've turned my attention to her — she'll open her eyes and gaze back — and I often revel in the sense of comfort this gives me, the two of us together in our quiet affinity. I almost never feel lonely when I'm in the company of the dog, and acknowledging this has been enormously instructive: she has helped me not only to understand the difference between solitude and isolation, but to live it. Out in the world with her, I have found a path to others. At home with her, I have found a way to be alone without the ache.

The dog-as-surrogate view implies that there are only two ways to inhabit the world, with other humans or without them, and it ignores the fact that sometimes you need both and sometimes you need a safe space somewhere in between. Dogs occupy that safe space; they make it possible.

9

THERAPY DOG

About a year after I got Lucille, I spent an afternoon with a woman named Miranda, a recovering alcoholic who told me about her first dog, a Doberman named Merlin. Merlin, she said, "was my first real relationship." She'd inherited the dog from a drug dealer on Beacon Hill while she herself was still drinking, and she described coming home one day about six weeks after she'd quit. She was living in chaos at the time, in a cold, barely furnished apartment in Dorchester with a crazed drug addict of a lover. The lover finally moved out, and on that particular day Miranda came home to find the few spare pieces of furniture gone and Merlin locked in a back room. He was so happy to see her, so relieved to be rescued from the enclosed room, and she sat down with him on the cold living-room floor and hugged him, and then she started to weep. She had felt utterly alone walking into that empty apartment, and she realized at that moment that in fact she wasn't. She held the dog, and she felt, perhaps for the first time in her life, that she was truly needed, truly responsible for another being, truly *in relation-ship* with another, and that awareness all but broke her heart. It also represented a tiny shift, as though this understanding set her a few paces down a new path: with that dog in her arms, she began to move away from her past and away from pain and toward a kind of comfort.

As I listened to Miranda tell that story, I remembered a conversation I'd once had with another dog owner, also a recovering alcoholic, who'd used the metaphor of an empty house to describe early sobriety. She said, "You take away alcohol, and you take away everything — your identity, your primary way of coping with the world — and it's like waking up in an empty house, absolutely nothing in it." She said this to me at a time when I was feeling fearful and bleak about life and hopeless about relationships — much the same way Miranda felt when she walked into her apartment that day — and I nodded into the phone. "Exactly," I said. "That's exactly how I feel, like my life is this big empty house with nothing in it, just me and a dog."

She paused. "Ah," she said, "but you've got that: this one beautiful, hugely important thing."

Yes. Merlin was this one beautiful, hugely

important thing to Miranda, and Lucille is this one beautiful, hugely important thing to me, and although I sometimes worry that she's part of the wall I've constructed — my excuse to stay *in here* and keep the world *out there* — I also know that her presence is what makes life behind that wall feel meaningful and rich. The dog is the one creature who can, and always does, penetrate the fortress. She swims across that psychic moat and trots right inside, and if I had to boil down what she brings me to one thing, it would be this: solace, a degree of healing.

The clinical literature on dogs as agents of healing is vast. Boris Levinson, an American child psychiatrist, coined the phrase pet therapy in 1964, following observations he made when he began to use his dog, a shaggy creature named Jingles, in sessions with severely withdrawn children. The dog, Levinson noted, served as an ice-breaker, softening the children's defenses and providing a focus for communication; with the animal present, Levinson could join in, establish a rapport, and begin therapy. Levinson wasn't the first scientist to study the use of animals in treating psychological disorders — interest in the subject dates back to the early twentieth century — but he was the first to write seriously and extensively about it, and he's credited with sparking widespread research into the phenomenon.

Scientists and health care professionals have since put Levinson's theories into practice in scores of therapeutic settings, and their results are uniformly consistent: animals can improve morale and communication, bolster self-confidence and self-esteem, increase quality of life. Psychiatrists Sam and Elizabeth Corson, two of the first to expand on Levinson's work, implemented the first pet-facilitated therapy program at a psychiatric unit at Ohio State University in 1977. In their study, fifty patients were allowed to choose a dog from a nearby kennel and interact with it daily at appointed hours. Three patients withdrew from the program; the remaining forty-seven showed marked improvement: the dogs acted as a social catalyst, forging a positive link between patients and staff; patients reported increased self-respect, independence, and confidence. A 1981 study in Melbourne, Australia, the first formal pet therapy program in that country, evaluated the influence of pets on morale and happiness among nursing home residents. Six months after the arrival of a former guide dog in the ward, a golden retriever named Honey, the sixty residents were rated as happier, more alert and responsive; they smiled and laughed more often and displayed more optimism about life. Members of a control group with no contact with dogs were less relaxed, more withdrawn, and less interested in others.

Study after study has supported such findings. Depressed patients in nursing homes have become more interactive and optimistic when visited by dogs and cats; prison inmates allowed to take care of birds and small animals have become less isolated, less violent, more responsible, and have exhibited increased morale (the pet therapy program implemented in a prison, at the Lima State Hospital, in Lima, Ohio, in 1975, has become a national model); visits by dogs and cats have helped ease feelings of fear, despair, loneliness, and isolation among terminally ill cancer patients; the presence of a dog at a psychiatric halfway house has helped residents become more social and more adept at communicating; elderly veterans, emotionally disturbed and learning-disabled children, and troubled inner city kids all have benefited from the presence of animals, becoming by turns more responsive and optimistic, more communicative and responsible, more compassionate.

Jack Stephens, a fifty-one-year-old veterinarian who established and now runs the VPI Insurance Group, in Anaheim, California, the nation's oldest and largest health insurance company for pets, understands the therapeutic value of pets, but he learned about it the hard way, through personal experience. As a practicing vet, Stephens had been aware of the rising tide of articles about the psychological value of pets in veterinary journals for years, but he'd always felt a tad skeptical about the concept, a feeling that crept into his view of his practice as well. People seemed so . . . over the top about dogs. He'd see owners indulging their dogs, crying over them when they got sick, dripping with emotion, and he just didn't quite get it; although he loved animals himself and always appreciated their company, he'd shake his head at half his clients, then go home and make little jokes about how overinvolved they were. "I think a lot of people in the veterinary profession don't quite realize how bonded people are to their dogs," he says. "There's still the idea that they don't have that much value."

Jack Stephens's own doubts about the bond disappeared about seven years ago, when he was diagnosed with throat cancer and underwent five grueling months of radiation and chemotherapy treatments. Several months before his illness was discovered, Jack's wife had acquired a miniature Doberman pinscher — another source of skepticism to Jack, a rather hard-boiled fellow who favored big working breeds and found something distasteful and rather unmasculine about little dogs. But this one — a small, sleek-coated, and uncommonly intelligent creature named Spanky — had a particular brand of charm that struck Jack as special from the start. He was a very expressive dog — when the family left him at home alone, he'd literally toilet-paper the house in protest, grabbing a roll of tissue from the upstairs bathroom, then running down the hall and stairs with it, streaming paper the whole way — and his affinity for Jack, whom he adored, had a quality of insistence that broke down Jack's defenses somehow. Spanky decided early on that he liked Jack's pillow, and each night he'd slowly and persistently edge Jack's head off it, then burrow into the pillowcase and spend the night curled up inside. Jack doesn't have a very high tolerance for that kind of behavior ("canine nonsense," in his words), but he couldn't help himself: in

Spanky, he found it completely endearing.

When Jack developed cancer and began treatment, his relationship with the dog changed. Some nights Jack's nausea got so bad, he'd sleep in another room. Spanky would quietly accompany him, staying close through the night, watching him. He had an instinctive sense for Jack's needs — how much distance he required and how much contact — and although Jack had a large and supportive network of friends and family members, the dog provided what he calls a "crucial link" in his recovery. Spanky made him laugh. He motivated him to get up and take daily walks instead of languishing in bed. His presence — entertaining, attentive, utterly undemanding — helped him avoid self-pity. Jack Stephens, an Oklahoma native who prides himself on his machismo, found himself referring to this tiny dog as "my little angel man from heaven," and his attachment to Spanky taught him volumes about the clients he once privately chided in his practice. The dog, he says, allowed him "to experience a different world with pets," to appreciate the power of our connections with them.

Jack Stephens's experience speaks to the profound satisfactions of living with a dog, the therapeutic properties of their mere presence. Some of this is physical: a number of well-known studies have shown that petting a dog — in some cases, even being in the same room as a dog — has a calming effect on people, reducing blood pressure and heart rate. But there's also something psychically healing about being with dogs, and you don't have to be a cancer survivor — or a resident at a geriatric nursing home or a juvenile delinquent or a prisoner or a psychiatric patient — to appreciate the effect.

The term "unconditional love" comes up; it's inevitable. Ask ten people to tell you what's so special about relationships with dogs, what's comforting and healing about them, and at least nine of them will say it: "unconditional love." Dogs love us — or they appear to — in a purer, more accepting way than our spouses or our best friends or even our parents.

I don't disagree with that idea. I believe that the dog's capacity for attachment is somehow less cluttered than ours, and that dogs can be far more steadfast and less demanding than many humans — these are the very qualities that help make them so effective in therapeutic settings like nursing homes and hospitals. And yet I've always found the term "unconditional love" to be a little thin when it comes to humans and dogs, static and lacking somehow. For precisely that reason, Steve Zawistowski, senior vice president and science adviser at the American Society for the Prevention of Cruelty to Animals, says he is on a one-man campaign to eradicate the phrase. "It implies," he says, "that dogs don't need respect, that you can beat and kick them and they'll crawl back." It also implies that the relationship is essentially nonreciprocal, as though our only role with dogs is to stand there and absorb. In fact, I think the healing power of dogs has less to do with what they give us than what they bring out in us, with what their presence allows us to feel and experience.

As Jack Stephens discovered, one thing they allow us to feel is closeness without claustrophobia, a sensation that's akin to unconditional love but is actually a little more complex — and a little easier to take. If Jack's wife, for instance, had been as attentive to him during his illness as Spanky was — if she'd sat up nights and watched him as he slept, perched next to him as he threw up in the night, never left him alone for a minute — he probably would have gone crazy, felt hemmed in, unable to breathe.

This may be true for all of us. The idea that someone will love us to death no matter what is a great and powerful universal fantasy, but it's also a bit of an impossible dream, one that can't — and probably shouldn't — be realized between grown men and women. We might talk a good game about unconditional love, about wanting it and not getting it from humans, but imagine a spouse who acted like your dog, who woke up every morning writhing with joy at the mere sight of you, who jumped up and down every time you walked into a room, who never uttered a critical word, who never took you to task for being irritable or neurotic or lazy, who gave you all of the power. In theory, that might sound fabulous; in fact, behavior like that could drive you around the bend in about five minutes. But dogs can give us what people can't. Perhaps because they're members of a different species, and so the line between them and us is clearly delineated, a dog can love you like that without raising questions of fairness, without triggering some confusing or destructive imbalance of power, without making you squirm. This is part of what can make our relationships with them feel so profoundly gratifying: dogs occupy the niche between our fantasies about intimacy and our more practical, realistic needs in relation to others, our needs for boundaries and autonomy and distance.

Within that niche life can be enormously comforting. Every night in the summer my friend Grace takes Oakley downstairs and lets her out into the yard. While Oakley pees and takes a moment to sniff around, Grace sits on the porch steps and looks up at the stars. When she's ready, Oakley trots up the stairs and sits down right next to Grace, and they stay there for a few minutes, gazing out into the night together. After a minute Grace will reach out and scratch Oakley's chest; without even turning to look at her, Oakley will hook her front paw over Grace's arm, and they will remain like that, a dog-human variation on a couple holding hands, until Grace decides it's time to head back upstairs. There is the niche, a perfect snapshot of it: one human and one dog in that unspoken and abiding attachment. "She's just right *there*," Grace says of Oakley, and that sense of presence, that sense that the dog is present *to Grace*, in a way she is not present to any other being on the planet, is what gives the connection its quality of singularity and meaning. Here I am with my dog. Me and my dog. The closeness feels like a private bridge, extending from human to animal, a causeway that nobody else can cross in quite the same way you do.

That causeway is constructed of ritual and repetition and simple moments, of behaviors discovered and then executed exclusively between human and dog, and there is something exceptionally restorative about crossing it day after day. A woman named Michelle, owner of a tiny white Maltese named Carmen, spends about thirty minutes every morning reading the paper in her kitchen; at some point her dog developed the habit of jumping up onto her lap, then snuggling underneath her nightshirt while she sips her coffee, and now this is standard practice, a moment of quiet, shared contact that never fails to please her. Michelle says, "It's this tender moment just between the two of us, before the kids get up and the phone starts ringing and all hell breaks loose. It's like our little special thing." A man named Andrew wakes up every morning to the sight of his yellow Lab's nose: "When she thinks it's time to get up, she walks over to the bed and just sticks her nose next to my head — inches away from me — and just sort of breathes on me until I open my eyes." The sight of this cheers him up every single morning. "That big, happy dog head," he says. "I swear I wake up smiling every day." For Nancy, owner of the German shorthaired pointer named Tomato, a particular kind of pet therapy comes in New York's Central Park, where she and her dog head every morning at eight o'clock to walk. "It is a very exquisite pleasure," she says. "Even in the worst weather, even in sleet. There are moments I feel like this is the last thing I want to do, and then I get there and I feel it's a gift to be out there. As much as I don't believe in God, I'm sure when I see the light coming through the clouds, I see God out there. Which I wouldn't do if I didn't have this dog."

These are all snapshots of that niche we inhabit with dogs, a place where the canine experience somehow intersects with ours, offering in the process a measure of solace. When I walk with Lucille, I'm not exactly sure I see God in clouds, but I do share Nancy's sentiment about the experience, the sense of comfort behind it. Out in the woods Lucille and I will sometimes stop and sit in the shade of a tree, and I'll watch her for a moment or two, watch the way her whole body is attuned to what's happening that very instant — what smells are wafting her way, what sounds she's picking up, what insect is flitting across her field of vision —

and I'll think: Ah, *right now;* that's what it's like to live right now. There's something very Zen about the experience of being with a dog in such a setting, something about the dog's orientation in the tangible here and now that rubs off on you, cues you in to sounds and smells and sights, eases temporal human concerns — what happened two hours or two months ago, what's going to happen two hours or two months later — into the background. Focus shifts; perspective returns. You find yourself exercising two skills that are so elusive in the human world: the ability to live in the present, and the ability to share silence. Snapshot of peace.

Next, train your camera on something more specific: the dog's coat, his fur. "Dogs are so *tactile.*" My friend Tom says this one day in his kitchen as he reaches down and scratches Cody, his Australian shepherd, and I think about Michelle with her Maltese, Andrew with his yellow Lab, Grace on the porch steps with her malamute. I think about the importance of touch, and about how nettlesome it can be in the human world, and I look at Tom's dog. Cody is the kind of creature who lives to be touched: if you're sitting near him, he'll often come up, crane his neck toward you, and just hold his head a few inches above your lap, a look of the most earnest expectancy in his expression, as though he's presenting himself to you like a gift. If you start to pet him, the look will shift from anticipation to gratitude and then bliss: ears back, mouth open in a smile, tiny stub from his docked tail in a nonstop wiggle. His whole body seems to say, "Pet me! Thank you! This is *good!*" and it can feel both startling and curative to be on the receiving end of that message. Our own culture is so dubious and suspect about touch, so rule-bound and witholding. Imagine reaching out to scratch a colleague behind the ears every time the impulse for physical contact hits you; imagine going up to friends and randomly hugging them or tousling ther hair. Behavior that would alienate humans (or get you arrested) is perfectly permissible with dogs, welcomed and solicited. Our need to self-censor vanishes in their company; touch is at our whim and need not be negotiated; in a sense, we are allowed to inhabit the physical world the way they do: openly, without self-consciousness.

This appears to be true across genders: in clinical observations of clients in the waiting room of a veterinary clinic, researchers at the University of Pennsylvania discovered that both men and women petted their dogs for similar lengths of time and with little difference in frequency. Their gestures were consistent, too: similar numbers of men and women stroked, scratched, and massaged their dogs, or rested a hand on the animal's neck, or sat with their arm around the dog, leading the researchers to conclude that dogs may have a particularly powerful role for men, especially those who tend to confuse touch with sexual overtures or possessiveness: "A pet may be the only being that a man, trained in the macho code, can touch with affection."

Alan Beck and Aaron Katcher suggest that the component of touch — our ability to touch dogs and their ability to touch us — gives the relationship between human and dog a quality of therapeutic intimacy, one that's both like and unlike the kind you might find with a traditional therapist. "A Rogerian analyst," they write, referring to the nondirective approach first advanced by psychiatrist Carl Rogers, "is not unlike a Labrador retriever." Indeed, the parallels between analytical and canine behavior are striking. Like an analyst, a Lab will not guide your conversation. He will not offer opinions or criticisms or tell you what to do; instead, he will be attentive but silent, observing you with an empathic gaze. Author Jerome K. Jerome put it slightly differently: "Dogs," he wrote, "never talk about themselves but listen to you while you talk about yourself, and keep up an appearance of being interested in the conversation." The big difference between dogs and therapists, of course, is that the dog can jump up and lick you, nuzzle you with his snout, let you kiss and hug him anytime the impulse strikes. "The difficult art in therapy," write Beck and Katcher, "is achieving a mutual feeling of intimacy without touching." With a dog, this is a piece of cake.

Another big difference between the Lab and the analyst: it's a lot easier to figure out what the retriever is thinking. My trainer, Kathy de Natale, lives outside of Boston with her husband and two German shepherd dogs. When I ask her why she's so drawn to dogs, what makes living and working with them so gratifying, she answers in a word: "Honesty. I love the honesty of the dog." Kathy herself is a very down-to-earth and straightforward woman, neither gooey nor sentimental when it comes to dogs, so she means this in the most literal way: with a dog, what you see is what you get. "Relationships with people are never so simple," she says. "There's so much to get through as a human that by the time we get to be adults, we're all carrying around all this baggage. But with a dog, you don't have to wade through a lot of garbage to get to what's real." I nod, listening to her, and think about how relieving that degree of clarity can be. Lucille wouldn't know a repressed emotion if it hit her on the head like an errant Frisbee. She wants her dinner, she sits on the kitchen floor and stares at her bowl. She wants a rawhide chew stick, she goes over to the drawer where she knows they are kept and she sits and stares at that. The dog's agenda is simple, fathomable, overt: *I want.* I want to go out, come in, eat something, lie here, play with that, kiss you. There are no ulterior motives with a dog, no mind games, no second-guessing, no complicated negotiations or bargains, and no guilt trips or grudges if a request is denied. If you've spent a lifetime navigating the landscape of human relationships, characterized as it can be by covertness and ambivalence and indirection, this can be an enormous relief.

So can the dog's lack of self-consciousness. Aaron Katcher has a wonderful childhood memory of being in his grandparents' living room, the staid and sober adults engaged in an elevated discussion about morality and religion while the family's Great Dane lay by the fire calmly licking his genitals. This is the kind of canine behavior that makes some people squirm, at least publicly: dogs, at least by our standards, can be so shameless. They lick and hump and sniff each other; they relieve themselves in public; they blithely act out all our repressed instincts and appetites; they are, Beck and Katcher point out, like four-legged embodiments of the human id. But if we get a bit embarrassed when a dog starts sniffing another dog's butt or mounting a visitor's thigh, I think we also get a certain vicarious satisfaction out of their freedom, their capacity to so freely flout our conventions. "We are drawn to dogs," wrote George Bird Evans in *Troubles with Bird Dogs*, "because they are the uninhibited creatures we might be if we weren't certain we knew better."

Sometimes that lack of inhibition rubs off on us: the dog's freedom becomes our freedom. "I *dance* in front of my dog." A woman named Linda, forty-five, tells me this over the phone in a breathless voice, then adds, "You have to understand: I don't dance. I don't think I've danced in public since junior high. I'm like the most self-conscious person on the planet." Her dog, a scruffy black terrier mix named

Cookie, is the one creature with whom Linda's reserve crumbles. "That's what I love so much about him," she says. "I can act like a raving lunatic in front of him, and I just don't think about it."

About a year ago I met with a blind woman named Barbara, thirty-seven, who lives with a guide dog named Homer, a two-year-old yellow Lab, and I expected her to describe slightly different sources of comfort: the particular levels of trust and communication you develop with a dog you depend on; the sense of partnership. She touched on those subjects — the two of them are "a team"; Homer is a critical symbol of mobility to her — but as she sat on her living-room floor with the dog and talked about their relationship, she returned far more often to the themes of simple joys: the pleasure of living with such an accepting and uncomplicated creature, the sweet earnestness of the dog, the freedom she feels in his presence to be herself. "It's so hard to talk about dogs without using clichés," she said, "but they really are best friends."

Actually, that's another sentiment I'd alter: in human form, even the best of friends may clash, disappoint one another, or grow apart, and nearly all close adult relationships require periodic tune-ups and reassessments. Not so with the dog. "The thing I love best about my dog," a woman once told me, "is that she will never walk into a room and say, 'Honey, I need some space.' " Amen. There will be no couples therapy for you and your dog, no three A.M. dialogues about who's angry and why, no relational bombshells at the breakfast table. In a world of social unpredictability, where you never know who's going to walk out on you or move to a new city or veer off in some unexpected direction, this is another great relief: the dog will never wake up one day and tell you he's been rethinking his whole life, needs some time off, wants a change, or wants to change *you*.

Over time the snapshots accrue; an album of constancy is amassed. Last spring I spent a few hours with a woman named Ann, who showed me pictures of her dog Claude, a standard poodle who lived to the age of sixteen, and died when Ann was fifty-six. The photos were a small testimony to the dog's unwavering presence in the midst of major life changes: there are Ann and Claude in 1965, standing in front of a moving van, poised to move from Chicago to Boston. There they are in 1968, right around the time of Ann's divorce. There they are in

1973, the day of Ann's father's funeral. There they are in her living room in 1975 — Ann is recovering from chemotherapy treatments, her head swathed in a scarf, the dog by her side. Ann shook her head, paging through the album. "So many changes," she said, "and that dog was right with me through all of them."

I have lived with a dog for only a heartbeat by comparison, but already I know what she means. Every night Lucille pads up the stairs behind me, then creeps rather shyly into the bedroom and sits by the bed until I invite her to jump up. This is the start of an evening ritual that never fails to please me, at least in part because it's so invariable. Lucille is at her sweetest and most polite at this time of day. She jumps onto the bed and lies down, head flattened against the blanket between her front paws, ears back, eyes shining. I get under the covers, and she observes me with an expression of delight and mild surprise, as though she's amazed that we're both here *yet again,* repeating this same behavior. Then I pat the covers by my hip and say, "Come here, you," and she squirms up next to me, pressing herself against the length of my legs. We spend several minutes together like this, the two of us in our niche. I scratch the dog's chest and shoulders; she stops me periodically, hooks a paw over my wrist and spends a moment or two licking my hand. After a while I'll pick up a book and begin to read; she'll lay her head down on my lap then and heave a great happy sigh.

Lucille and I relive this scene every night, and I'm often struck by the fact that no matter what kind of day I've had — good, bad, indifferent — it always ends with this, this tender sameness. My human relationships are unpredictable, sometimes volatile, always subject to complication and flux. But my dog stays the same, her reactions to me constant. In a sea of changeable emotions and circumstances, she is a small anchor, a steady presence who bears witness to the most private details, the monumental shifts and incremental changes, who remains right there.

I lie there and I pet her and I breathe.

"This is the only nonpolitical relationship I've ever had," Ann said, a comment that struck me as odd and also perfect. Dogs represent the one relationship in life where consistency is never questioned, never doubted, never compromised by the vicissitudes of human moods and circumstances and priorities. And so we get to experience something else that's rare in human affairs: trust.

Talk about healing. Dogs are there with us, as Jack Stephens found, the way we wish our spouses were, for better and for worse, in sickness and in health.

They may even be with us in death. Jonathan, the self-described codependent owner of the basenji named Toby, lost his lover, Kevin, to AIDS in 1994. In the last weeks of his illness, Kevin was nearly comatose, suffering from an extremely high fever and unable to maintain consciousness for more than a few minutes at a time. Toby, then a little more than a year old, maintained the most careful vigil over him. For the full course of his illness, Toby would lie on Kevin's chest like a cat, or curl up tightly against his feet, and he'd growl whenever anyone approached, even Jonathan; if Jonathan tried to pull him off the bed, Toby would resist and hurl himself back.

Kevin was hospitalized the night before he died, and as he neared death, he began to call out to the dog, who of course was not physically present in the room. "Toby, stop chewing that!" he'd say, or "Stay away from there, Toby!" Or he'd lift up the covers on his hospital bed and appear to be letting the dog under the sheets. Jonathan says, "Kevin had this funny little thing of blowing on the dog's face, and Toby would do this fly-swatting thing in response, batting him in the face. In the hospital Kevin was doing all these things — blowing in the air, smiling. I was on one side of him, and our friend Susan was on the other side, and this imaginary dog was in the middle. And at the very end, Kevin said, 'Toby! Toby!' And he was kissing the air, and then he died."

There is no doubt in Jonathan's mind that Toby was there with Kevin in some spiritual sense, that the dog made his presence felt, that he helped him to die.

In turn, Toby helped Jonathan to live. Jonathan was stunned with grief, but there was this animal, Kevin's beloved dog, serving as a connection to the person he'd loved. There was this animal, who needed to be fed and attended to, who'd bat his paws in the air at him when he wanted to be taken out for a walk. There was this dog, leading him out into the world, helping him navigate the landscape of loss. "After Kevin's death," Jonathan says, "Toby was the only thing I wanted to be with. After about six months I did this grief workshop, and they asked us to write down all the things that were really important in our lives. I couldn't think of one person to put down, but the one thing I could say, unconditionally, was Toby."

★ ★ ★

Dogs can help us heal past wounds as well
as present ones. My friend Mary and I,
who've both racked up many years of
struggle in psychotherapy, talk a lot about
how hard it is to move past old sources of
pain, about how you can spend eons with a
therapist and relive in the most wrenching
way the deepest disappointments and hurts,
and how in spite of that kind of work, the
distress never fully abates, the voids never
really fill. You think you really understand
the ways your mother let you down, understand finally and fully the ways you felt mis-
understood or undervalued or disregarded
as a kid; you think you're past all that at last,
and then you go shopping with your mom,
or walk into her house for dinner, and you
regress in about ten seconds, one minor empathic failure tapping in to an ocean of
older, larger ones.

Mary has a four-year-old Chesapeake Bay
retriever named Georgia, a beautiful female
with a soft, curly brown coat and large
greenish-yellow eyes. Out to dinner at a
restaurant recently, Mary's mother kept
referring to the dog as "he." "How is he
doing?" "Does he shed a lot?" Not a huge
transgression in the grand scheme of things,
but her mother's refusal to remember the

one simple fact of the dog's gender felt like a slap; it opened all those psychic memory banks, harkened back to countless episodes, big and small, that had left Mary feeling disconnected and minimized in the past. *She doesn't get it.* That's how she felt, for the nth time. *My mother doesn't get the things that are most deeply important to me; my mother doesn't get* me. She sat there and she clenched her teeth and she wanted to scream.

But this is the thing: Mary got home that night, and as she pulled into the driveway, she heard her dog — her big, beloved female dog — barking from the upstairs window, anticipating her arrival. She headed up the stairs, and she heard Georgia's toenails scratching against the floor behind the door, heard her high expectant whine: You're home! You're back! She opened up the door, and she felt sixty-two pounds of retriever hurl itself into her arms, and she sat down in the hallway with that dog, and she felt loved and needed and valued, and she loved that dog right back, sat there and rubbed the dog's belly and scritched her ears and giggled and cooed. You can't fully transcend old pains, can't inure yourself completely against the times when they emerge from their buried places, but you can find new ways to soothe them, to ease the blows. Mary can love that dog in a way she herself was never loved, and — just as amazing — the dog can love her back. "This dog loves *me,*" she says, and she shakes her head in wonder at that simple notion, at how comforting the knowledge is, how much it fills her up.

You hear this a lot among people who have very intense relationships with their dogs: the dog offers a kind of corrective emotional experience, allows us to both give and receive what we haven't quite gotten in our human relationships. Sometimes dogs are the mothers we always wanted, sometimes they are the children we never quite got to be, often they are both. Kathleen, the woman who's been known to prepare a special plate of pancakes for Oz, her Australian shepherd, grew up in the most chaotic household. Her father was a jazz musician who worked nights, her mother was a registered nurse who worked days, and both were needy and rather childlike people who really didn't have a clue how to raise a kid. Kathleen went into day care when she was only a year old; her mother used to joke that Kathleen "raised herself"; and although her parents were basically well-intentioned, kind people, Kathleen never had the sense that they were looking out for her, that anyone in the world really put her first.

Today Kathleen lives with her own share of chaos — she's divorced, a single mother who struggles to make ends meet — but she also has this dog, Oz, this wonderful creature who literally stands up beside her at the kitchen counter when she cooks dinner, his front paws on the butcher block. She has given Oz an *un*chaotic place to live, a lifestyle that's simple and predictable and calm. She has loved him with an eye toward his needs: when Oz was a puppy, Kathleen worried about stifling his shepherd instincts, his need to herd, and so she used to take him to a park near her house and pretend to be a sheep. She'd stand there and say, "Baaaa! Baaaa!" and then she'd *run,* and Oz would tear around and chase her. That purity of love you give a dog — a feeling we all long for in relationships and so rarely experience — has not wavered since she got Oz, and Kathleen finds this deeply satisfying, this ability to love another being in such an uncomplicated way. And the receiving is as corrective as the giving. Dogs, after all, do resemble ideal mothers in some important respects: they're totally interested in us, totally accepting, fascinated by just about everything we do, and this reality is not lost on Kathleen. Oz is as much her parent as she is his, more attuned to her than her real parents were, far more available and consistent. "He totally comforts me," she says. "And I don't think I've ever had a relationship in my life that's been a comfort."

Dogs allow us to rewrite the childhood script. Emily, a thirty-six-year-old woman who was badly neglected and abused as a child, has created with her dog a model of care that eluded her as a girl. The dog, a chocolate Lab named Riley, has a good diet, regular exercise and medical care, the very cleanest ears and eyes and teeth, and nurturing him this way has been life-saving for Emily in the most literal sense. Her upbringing left her with a dissociative disorder and a long history of depression, and as she tells me, through tears, "Riley is what keeps me sane. Suicide is always on my mind, and I can't do it — I know I just will not do it — if I have Riley. I can't abandon him like I was abandoned. So it's sort of like how I stay alive is because of my dog. He keeps me going."

Debbie, forty-nine, has invented perfect siblings of sorts. The middle child of three girls, she was the family caretaker as a kid, the great mediator, and fairness has always been, she says, "a huge issue" for her. When she talks to me about growing up, she uses the phrase "game warden" several times: "I was the game warden for the three of us"; "That was my role: game warden." Today, she lives with three female Rhodesian ridgebacks, and the phrase comes up again when she talks about them. "There are three of them, always vying for attention or fighting over a bone, so in a sense I'm always having to be the game warden." This transfer — mediating dogs instead of sisters — has been marvelously effective. Debbie's role as family healer backfired in human form (when she finally stopped taking care of everybody else and started putting her own needs first, she ended up estranged from both sisters), but the dogs have allowed her to recreate her early family experience without the attendant costs. Debbie understands the canine hierarchy — which of the three dogs is the alpha, which comes second, and which comes third. She respects the pack system, knows that the dogs will be happiest if she comes into a room and pets the alpha dog first, ignoring the other two. She knows how to keep the peace, and the act of doing this — of ensuring that everyone is treated fairly, that everyone's needs are met — is as satisfying to her as the act of attending to Lucille with such care is to me; it hits some central note of familiarity, answers the oldest longing.

What's funny about these historical revisions is the elusive, serendipitous way they take place. Kathleen and Emily didn't set out to recreate their own upbringings through their dogs; Debbie never sat up one day and thought: Well, gee, I wonder if I could be a different kind of game warden if I were working with three female dogs as opposed to three female humans. It just sort of . . . happened. I hear these stories, and I'm reminded of the way I stumbled toward Lucille, of the semiconscious way I managed to acquire not only *a* dog but *this* dog. All dogs can be guide dogs of a sort, leading us to places we didn't even know we needed or wanted to go.

Sometimes a dog can lead you to several different places at once.

Consider a scenario:

The dog and I are playing soccer. We are deep in the woods, and I am standing in the middle of a trail, a blue ball under my foot.

"Ready?" I say. "Are you *ready?*"

Lucille leaps ahead, dashes about four feet in front of me, and plants herself in the middle of the path, facing me. Then she crouches into a play bow, her whole body tense with concentration. She watches my foot, waits. I see her paws tighten against the dirt path, see her tail sweep up behind her. I step back, give the blue ball a swift kick in her direction, and she pounces on it, the goalie in action. Sometimes she stops the ball as it comes toward her, scoops it up in her mouth, and prances off, great pride in her step, but usually the ball flies past her and she has to dash off in pursuit. Several times the game has ended in near disaster. This particular ball — small, with a bell in it — is of great importance to Lucille: it is the only ball she will play with, the only ball she ever loved, and it appears to be irreplaceable. Throw a substitute ball, even an identical one, and she will sniff it disdainfully and walk away. So I live in fear of losing the blue ball. I have also inadvertently kicked it into barely retrievable places and have had to mount heroic rescues to save it, wading through pond scum to pick it out of a marsh, plunging my arm, shoulder high, into December waters to pluck it out of a lake, even scaling a seven-foot chain-link fence to retrieve it when it disappeared through a hole. But no matter; the effort pays off. Armed with the blue ball, Lucille gets to run, and I get to act like a kid. I shout — "Good save! *Yes!*" — and I run after her, and I tear off my jacket and scale fences, and I laugh out loud.

I also get to slide in and out of about six other personae. Sometimes I am the dog's playmate; other times I am her coach, enthusing over a successful catch or a well-executed command; still others, I am the disciplinarian, tramping through the brush to retrieve her if she refuses to come when I call (yes, this still happens). The dog can feel like a sister, or a best friend, or a protective older brother, she can feel like a mother or a child or both at once, and there is a quality of effortlessness to the way we adapt and exchange these varied roles, as though we've both learned how to doff Hat A and replace it with Hat B without so much as a flicker of negotiation.

I suppose this is my own haphazard form of pet therapy, this steady dipping in and out of different sides of the self, and on balance I'd say it's been effective: little pieces of me, some of them long dormant, have been cultivated, crystallized, comforted in the process.

I have relearned something about play on those walks, something that got drained out of me for a while in all those years of drinking and grief. Sometimes this hits me as an awareness of absences: I'll walk along beside Lucille and realize — oh! — I'm not clenching my teeth, not stiff with tension, not chewing over a worry, and this is so different from my prior state, I barely recognize the sensation, which is ease. I also smile a lot more than I used to. Last fall I scheduled a session at my therapist's home office so that I could bring Lucille, whom he'd never met. My therapist's dog, a charming, seven-year-old West Highland white terrier, joined us, and Lucille, who has a very vivid enthusiasm for terriers (they bring out the devil in her), appeared to consider this a fifty-minute play session in her honor. She was rapt: she wrestled the terrier, she pawed at him, she threw herself on the floor on her back in front of him; to my horror, she got up on her hind legs and humped him. The hour was in some respects useless — we gave up trying to talk about anything but dogs after about seven minutes — but I found myself doing something I hadn't done in thirteen years of psychotherapy, which is laugh through an entire session.

Playmate, soulmate. My relationship with Lucille has also touched the twin in me, providing a sense of exclusivity and interdependence with another being that I haven't experienced since childhood, growing up with my sister. This is something I hadn't quite realized I'd lost, or even missed, until I got the dog — a depth and singularity of attachment, a sense of being truly entwined with another being — and I'm often aware of its return when we're out walking or hanging out on the sofa together at home. *Us,* that pack of two. I'm more important to her than any other human; she's more important to me than any other dog. I suppose this is the way my sister and I felt about one another as kids, before we grew up and began the long, arduous process of becoming distinct individuals, and Lucille has sort of picked up that thread of feeling, reintroduced me to it. She and I are often together, it's worth noting, in many of the same ways my sister and I were together when we were young — constant companions, playmates, bedmates — and I'm struck by how familiar that brand of proximity feels. I wouldn't want to be that bonded with an adult human, that inseparable or entwined, but I relish being able to tap in to the sensation with Lucille.

I imagine lots of us do, consciously or not. Constance Perin, a cultural anthropologist, suggests that the depth of our affection for dogs has to do with precisely that dynamic, with the dog's ability to help us recreate a brand of closeness we experienced with our mothers as children. Elaborating on Perin's ideas, psychiatrist Eleanora M. Woloy writes, "In no human relationship can we recreate this symbolic vessel of excess love. . . . The relationship [with the dog] becomes the memory of the hope for a magical once-in-a-lifetime bond."

This, she suggests, is what gives the relationship its elusive quality of resonance: there is something deeply familiar to many of us about the singularity of a dog's affection, and something deeply healing about tapping in to the kind of closeness they provide. Loving — and being loved by — this attentive, present creature helps heal the disappointment we felt as kids, when we became aware that we had to cope with the world on our own; it mitigates one of our most primal struggles, between the wish to merge with another and the need to separate.

And, as it is wont to do, the healing helps, trickles down into our other bonds. I can't say Lucille is directly responsible for this, but since I've had her, and had a chance to bask in the familiar warmth of that old exclusivity, my relationship with my sister has lost some of its more recent strain; my quiet sense of resentment toward her, for going off and leading her own life, has abated; I feel closer to her but in a cleaner, less entangled way than I used to, as though the idea that we're separate individuals with our own needs and priorities finally makes sense. In the spring of 1997 my sister acquired her own dog — a lovely, genial, four-year-old border collie–black Lab mix named Beamer — and we meet for walks a couple of times a month, outings that are a source of great comfort to me in and of themselves. Unexpected bonus: the dog has restored the twin in me in both real and metaphorical ways.

So there we are: one human as friend, child, mother, twin; one dog as dog, willing partner to them all. Put those varied combinations together, and what emerges, finally, is the most important role of all: human as human, a creature who's capable of love.

One night not long ago, I turned off the TV and got up from my chair in the living room, ready to rouse the dog from her station on the couch and head upstairs to bed. I looked at her. Lucille was lying on her back on the sofa — utterly ridiculous, the hind legs splayed out in either direction, neck craned to one side so that her head was all twisted around, snout pointing down toward her tail. She does this a lot and it kills me every time: she looks so silly, and also so profoundly vulnerable and exposed, forty-five pounds of absurdity and trust. We regarded each other for a moment, me standing above her, the dog peering up at me from this ludicrous position, and then I crouched down by the sofa and started to rub her belly. And then I just melted, some piece of me inside just melted at the sight of her, and I heard myself say aloud, "I love you every single day. *Every single day.*" This sounds so corny, cooing words of love to a dog like that, but even as I heard myself speak, I was aware that there was something miraculous about it, something miraculous and profoundly healing about the fact that I love this animal and find joy and solace in her presence 365 days a year, without exception. I have never felt that unwavering in my affection, never really felt safe enough to allow it. My human relationships have characteristically been about withholding — keeping parts of me shut down, or held back, or under wraps, protected against disappointment or vulnerability. My relationship with Lucille is about giving, an unrestrained, fearless, expressive kind of giving that's brand new to me, and it makes me feel human, it makes me feel whole.

I sat beside the dog for several minutes that night, petting her and looking around the room. I moved into my house four years ago at the most unsettled time: newly sober, still dazed with loss, my sense of self all foggy and my future so uncertain. I chose to buy a house at that time because I needed a way station, a place to rest until the dust from all that death and drink began to settle, and I've always loved the structure itself, a small Victorian with high ceilings and odd angles and lots of light. But until the dog came, the house felt in some ways as un-formed and empty as I did, and I'd find myself walking around inside it like a visitor at a museum, thinking: Nice place, but do I really *live* here? Today the living room sometimes looks more like Lucille's place than my own — dog bed in front of the sofa where a coffee table might be, box of toys on the floor where an end table might be — but that early sense of unfamiliarity has ebbed away, as though the rough, unsettled edges of my own soul have grown smoother. I looked down at the dog, this quiet picture of acceptance and contentment on the sofa beside me, and I thought: Home.

EPILOGUE

Shortly after I got Lucille I took her with me to Martha's Vineyard, to my family's summer home. This is loaded territory for me: it's the site of all those long summers I spent with my family, feeling like an outsider, so restless and bored; it's also the site of a lot of ghosts. My father's ashes are buried some distance from the house, in a grove of trees you have to wind several minutes through the woods to get to. My mother's ashes are closer: they're buried about twenty paces from the front porch, under a cherry tree, and my brother and sister and I put them there deliberately, a few feet away from the ashes of her last dog, Toby.

It sounds strange to say this out loud — we buried our mother's ashes next to the dog's, rather than her husband's — but the placement made sense at the time, and it still does. My parents' relationship fell apart in the last year of their marriage, and although we didn't articulate it quite this way, I think we all had the sense that her bond with the dog was purer somehow than her bond with my dad, more honest and more loving.

Toby died two years before my father did, quite suddenly, at the age of eleven. He was on the Vineyard at the time, and he'd been racing back and forth across the porch, chasing two of my nephews, sons of my half-brother who was up visiting for the weekend. It was midsummer, very hot, and my mother called out to the kids several times: Be careful with the dog. Elkhounds are not hot-weather dogs; they have thick double coats, and my mother was worried about overheating Toby. She heard a noise at one point. Toby gave out a great heave and collapsed, just collapsed, right there on the porch. My mother raced out, thinking he'd stumbled on something, tripped or hurt a paw. She got to his side, and he lifted his head and let out another eerie, gasping sound, and then he was gone. That afternoon she and my dad wrapped him in a small rug and drove him to the vet, a forty-five-minute trip to the other side of the island. An autopsy revealed that he'd had a heart attack, and we later speculated that he might have had Lyme disease: he'd been kind of achy in the weeks before his death, his joints apparently stiff, which is one of the symptoms; Lyme disease can weaken muscles, including the heart.

My mother was not a weeper. My whole life I saw her cry only twice, the first time when I was sixteen and she had to euthanize her first dog, Tom, who suffered from kidney disease, and the second time over Toby. She didn't cry, at least not publicly, when her own parents died; she didn't cry at my father's funeral. But when she called me up the night Toby died, I could hear her voice quavering. I don't remember much about our conversation, but I remember her shock and her sadness, the sound of her sniffling over the phone. She had Toby cremated, and she put his ashes in a little urn, and for an entire year she kept that urn on the floor of her bedroom, the spot where Toby used to sleep.

We buried the ashes the following summer. My father, several months into his illness, was confined to a wheelchair by then and couldn't join us, so we parked his chair by the porch window, and he watched me and my brother and sister and mother troop out to the cherry tree, where Toby used to like to lie and sniff at the air. We dug a hole and poured the ashes in, and my mother tried very hard, and unsuccessfully, not to cry. Her eyes brimmed and brimmed, and her face got all red, and a few big tears spilled out and ran down her cheeks. We poured the ashes into the hole and covered them over, placed some rocks and shells around the site. Then we stood there for a few moments, honoring my mother's sadness as much as the dog.

I couldn't have appreciated the quality of her loss at the time. I remember thinking: Yes, this is very sad, losing the dog, but I also remember speculating that some of her grief was deflected, that my mother was crying for Toby because it was easier than crying for my father, simpler somehow. I looked across the porch and saw his silhouette against the window, my dying father in his wheelchair, and I thought: There's the real sadness, it's for him, for their marriage.

I've come to believe I was wrong, that the sadness was, in fact, for Toby, and for the crater in her world left by his absence.

A year or so after I got Lucille, I had dinner with a friend, and I started to talk, somewhat casually, about a conversation I'd had with Carole Fudin, the New York psychotherapist who'd spoken to me about some of the bereavement counseling she's done with people who've lost pets. My friend sat bolt upright in her chair. "Bereavement counseling? For *pets?* Oh, Jesus." Then she laughed out loud. I sat there feeling a little stung, as though she'd slapped me, and I thought about my mother that day on the Vineyard, her tears. I thought about the way she'd sit at the kitchen table and scratch Toby, the one being in the household she felt free to touch. I thought about the constancy he'd provided her over the years, the sense of presence he'd offered this private and solitary woman. The one imperfect aspect of this near-perfect relationship, our bond with dogs, is its lack of longevity. They live such brief lives, and if there's one thing that intensifies the sense of loss we experience when they die, it's the fact that our grief tends to be pathologized, considered excessive and misdirected, even silly.

Last spring I spent a few hours with a couple who'd lost their dog, a beautiful, slender sled dog named Kimmi, in a freak accident in the White Mountains of New Hampshire. Kimmi was more like a spirit than a dog. I used to see her at Fresh Pond from time to time, and I'd watch her run: she had long legs and perfect proportions, a silvery gray coat, wide soulful eyes, and the most light-footed agile gait I've ever seen. People used to stop and stare at her the way they stop and stare at a beautiful woman, she was that unusual and that breathtaking. Her owners, a couple named Tom and Sue, went fly-fishing with Kimmi when she was just about a year old. They hiked alongside a river, and something happened — some trees came dislodged from the hillside along the riverbank, came avalanching down and hit Kimmi, breaking her spine. Tom and Sue spent hours trying to rescue her: Sue sat with her at the bank of the river in the most abject horror, and Tom raced off to get help, and when they talked to me about her death (Kimmi didn't survive) nearly a year later, they were still reeling from it, stunned by its depth. Tom, a psychotherapist, went back into therapy, his sadness was so profound. "I have seen death with patients, and I've had both my parents die, and I've had a friend that died, so I've been around death some. But I haven't grieved for anyone, not even my parents, the way I grieved for that dog," he said. "I don't normally cry, but I'd *sob*, and at night I'd wake up at two in the morning and — ugh, I can remember just waking up and gasping. And I've never done that for a human."

That's the sort of description only people who love their dogs deeply understand. I couldn't have understood it before I got Lucille, didn't understand it when my mother lost Toby, but I do understand it today. The loss is as particular and profound as the intimacy, and the depth of mourning it sets off can shock people, for we're often not fully aware of how many voids the dog has filled until he's no longer there, no longer filling those spaces in his able, silent way. When Marjorie lost her first border collie, Glen, she could not say his name for an entire year, literally could not utter the name Glen without bringing on a migraine headache. She told me that, and I just nodded: I can only imagine, I said. In fact, I can barely wrap my mind around the idea that I am likely to outlive Lucille, although a part of me lives in secret terror of the prospect all the time. "I've been worried about this dog's death since she was twelve weeks old." I say that to people a lot, and I'm only half kidding.

When Lucille was a puppy, I used to see a man named Charles at a local park every morning. Charles was an older guy, mid-fifties I'd guess, and he had a big old lumbering black Lab named Ben, and they'd stroll into the park together, Ben off-leash but glued to Charles's side, and hang around for a little while, ten or twenty minutes. Ben was all gray around the muzzle, as big and sturdy as a sofa, and he adored Charles: he'd stand there and gaze up at him, tail doing a slow, perfect wag.

Charles would give him a biscuit, and then they'd wander out of the park and head home.

Ben died this past March. I ran into Charles on the street several weeks later, just after an early spring snowstorm. He was standing near his front porch clutching a shovel, and he seemed grateful for the opportunity to talk about the loss, the way you're grateful after a human's death to run into someone who can appreciate how bereft you feel. Charles is a man of rather few words, but he talked about his sadness, and about what a special dog Ben had been, and then he pretty much summed up what life in the aftermath of his dog's death felt like, what it was like to reach instinctively for the leash and realize you no longer need it, to feel that empty space where the dog used to be: "It's like you have to reprogram your whole nervous system," he said.

I stood there, Lucille by my side, and my heart just went out to him. Loving a dog deeply does have a cellular quality, as though the most central part of you — your whole nervous system — gets tied into the bond, into the life you create together. You do get reprogrammed: a person with a dog becomes a dog person, with all the change that implies.

★ ★ ★

I suppose I've been aware of that from the beginning, the power of the dog to change who you are. Lucille was only about thirteen weeks old when I brought her on that trip to the Vineyard; I'd had her for only ten days, and I remember leading her out to the cherry tree on our first morning, to the site of those ashes. I watched her sniff around the two circles of stones, and I had the sense as I stood there that I was introducing her around: Mom, this is Lucille. Toby, Lucille. There was a quiet, sad feeling to this, and also a sense of uncertainty, as though I were standing on the edges of things unknown, a transition perhaps, or several transitions. I had not been to the Vineyard since the previous summer, when we buried my mother's ashes, and standing there with Lucille, I was jarred anew by the sense of how much had changed in such a short time. Both parents, gone. Drinking, gone. New life, new dog, sense of self still a blank.

So it was really *me* I was introducing around: Mom, here I am with a puppy. Toby, here's your successor. That was the sense I had: Here I am showing you the outlines of some new identity. Only I wasn't yet sure what that meant, what that identity would turn out to be, or what role Lucille would play in its development, the outlines felt so vague and ill defined.

Today, nearly three years later, the outlines are a little clearer, the picture emerging in sharper detail. There is still a lot to be filled in, a great deal of fuzziness around the outskirts: Is that woman in the picture a solitary person or an isolated one? What else will fill up that big empty house of early sobriety? Who else? But at the center I can see the clearest image, the most important one: a woman holding a leash instead of a drink; a woman with a dog by her side, *this* dog.

It's often said in recovery circles that it's very hard to give up an addiction without finding something to replace the loss, something that feeds you and fills you up and identifies you in some of the same ways the initial substance did. I'm too much of a romantic to see Lucille in such clinical terms, as a "replacement" for alcohol, but I can say that in loving her I have had that sense of being filled anew and essentially redirected, an old identity shattered and a new one emerging in its stead. A sense, as Charles might put it, of being reprogrammed.

Lucille has never seen me drunk. Simple enough statement, but it means volumes to me; she is a central part of the solace and peace I derive from being a sober person. I look at her sometimes at night, and I think about what a mess I'd be if I were still drinking, about how compromised my ability to care for her would be, about how she'd look at me and know that I wasn't really there. She is a symbol of something that was unavailable to me in the throes of addiction, an emblem not just of what I've been able to give to another being but of what I've been able to give to myself: consistency, continuity, connection. In a word, love.

Mystery novelist Susan Conant has written, "I feel the same spiritual comfort holding a leash that others feel holding a rosary." I know exactly what she means: put a leash in my hand, put Lucille by my side, and something happens, something magical; something clicks inside, as though some key piece of me, missing for years, has suddenly slid into place, and I know I'll be okay. The feel of that leash is as grounding and vital to me as a glass of white wine used to feel, the dog as central to my sense of well-being in the world as drink once was.

What makes you feel empty and what makes you feel full? Who, or what, makes you feel connected or soothed or joyful? How much companionship do you need, and how much solitude? What feels right, what feels like enough? We all have to feel our way through those questions in life, and although she cannot provide the answers for me, I have the sense that Lucille is gently leading me toward them. I pick up that leash; I go forward.

SOURCE NOTES

THE COLOR OF JOY
Page 16 Jean Schinto, author of *The Literary Dog*: "To deny dogs their nature is to do them great harm": From Winokur, Jon (ed.), *Mondo Canine* (New York: Dutton, 1991), p. 18.

24 More than one third of all Americans live with dogs today — that's about 55 million dogs: *Statistical Abstract of the United States: 1997*, US Census Bureau. See also: 1996–1997 *APPMA National Pet Owners Survey*, The American Pet Products Manufacturers Association, Inc. (Greenwich, CT), 1997.

25 The average owner can expect to shell out a minimum of $11,500 in the course of his dog's life: "Hey, Big Spenders," *The New York Times* (Sept 11, 1994).

25 Eighty-seven to 99 percent of dog

owners report that they see their dogs as family members: Cain, Ann, "A study of pets in the family system," in Sussman, M. B. (ed.), *Pets and the Family* (New York, Haworth Press, 1985), pp. 5–10 (87 percent of respondents in her study of 60 families reported seeing dogs as family members). See also: Voith, V. L., "The companion animal in the context of the family system," ibid., pp. 49–62 (99 percent of respondents in a study of 500 owners considered their pets to be family members); and Catanzaro, T. E., "The human- animal bond in military communities," in Anderson, R. K., Hart, B. L., & Hart, L. A. (eds.), *The Pet Connection* (Minneapolis: University of Minnesota Press, 1984), pp. 341–347 (98 percent of respondents in a study of 896 military families considered the pet a family member).

25 The American Animal Hospital Association annual surveys of pet owner attitudes: *National Survey of People and Pet Relationships*, American People and Pet Relationships, American Animal Hospital Association (Denver, CO), 1995, 1996.

Page 34 People exhibit a fair amount of species loyalty when it comes to acquiring pets: Serpell, James, "Childhood pets and their influence on adult attitudes," *Psychological Reports*, Vol. 49, pp. 651–4 (1981). See also: Kidd, A. H. & Kidd, R. M., "Factors in adult attitudes toward pets," *Psychological Reports*, Vol. 65, pp. 903–10 (1989); and Poresky, R. H., Jendrix, C., Mosier, J. E., & Samuelson, M. L., "Young children's companion-animal bonding and adults' pet attitudes: a retrospective study," *Psychological Reports*, Vol. 62, pp. 419–25 (1988).

41 "You ask of my companions. Hills, sir, and the sundown, and a dog large as myself that my father gave me": From Johnson, Thomas H. (ed.), *Emily Dickinson: Selected Letters* (Cambridge: Harvard University Press, 1996 edition), p. 172.

42 "Sagacious, vigilant, impressive, with all his faculties in a radiant intensification": Mann, Thomas, *A Man and His Dog* (New York: Alfred A. Knopf, 1930), p. 188.

42 "Topsy is my friend . . . dogs are children that do not grow up, that do not depart": Bonaparte, Marie, *Topsy: The Story of a Golden-Haired Chow* (New Brunswick, NJ: Transaction Publishers, 1994), p. 124.

NINETIES DOG

Page 72 Pet-owners who book their pets rooms in hotels that offer bone-shaped beds and special doggie room-service menus: "Active adults bring their pampered pooches along," *USA Today* (Aug. 20, 1996).

72 Manhattan dog owners who are spending up to $10,000 per year on their dogs: "A dog's luxe life," *The New York Times* (July 20, 1997).

72 A health-and-fitness center for dogs in Westwood, California, that features treadmills, Jacuzzis and swimming pools designed for dogs: "A new leash on life," *People* (April 28, 1997).

74 One quarter of the US population lives alone: *Current Population Survey*, US Census Bureau, March, 1997.

74 One in two marriages end in divorce: *Advance Report of Final Divorce Statistics*, 1989 and 1990, Vol. 43, No. 9, National Center for Health Statistics.

75 Twenty-one million women are divorced or single mothers: *Current Population Survey*, US Census Bureau, March, 1997 (divorce statistics); *Household and Family Characteristics*, US Census Bureau, March, 1994 (single mothers).

79 Easily half of today's dog owners name their dogs after people: *National Survey of People and Pet Relationships*, 1996.

79 The majority of today's dogs are allowed to sleep in their owners' bedrooms: ibid. See also: Voith, V. L., in Sussman, M. B. (ed.), *Pets and the Family*.

84 "The Kilcommons way to a perfect relationship": Kilcommons, Brian, with Wilson, Sarah, *Good Owners, Great Dogs* (New York: Warner Books, 1992).

84 "We approach training as a way of relating to your dog": The Monks of New Skete, *How to Be Your Dog's Best Friend* (Boston: Little, Brown and Company, 1978), p. 13.

84 "In a dog's mind, a master or a mistress to love, honor, and obey is an absolute necessity": Woodhouse, Barbara, *No Bad Dogs: The Woodhouse Way* (New York: Summit Books, 1978), p. 59.

87 By the mid 1950s, televisions were installed in the majority of American homes: Chafe, William H., *The Unfinished Journey: America Since World War II* (New York: Oxford University Press, 1986), p. 129.

BAD DOG

Page 113 Koehler refers to over-emotional owners as "wincers," "humanaics," and "cookie people": Lenehan, Michael, "Four ways to walk a dog," *The Atlantic Monthly* (April 1986).

123 Four to six million dogs are given up to shelters each year: Beck, Alan & Katcher, Aaron, *Between Pets and*

People: The Importance of Animal Companionship (West Lafayette, Indiana: Purdue University Press; revised edition, 1996), p. 202.

123 As many as 40 percent of dogs surrendered to shelters are given up by disillusioned or frustrated people who didn't realize how complicated living with a dog can be: Beck & Katcher, op. cit., p. 203. See also: Arkow, P. S. & Dow, S., "The ties that do not bind: A study of the human-animal bonds that fail," in Anderson, et. al., *The Pet Connection*, pp. 348–354.

123 About 25 percent of pets are destroyed by the time they reach two years of age: "Reigning cats and dogs," *Time* (August 16, 1993).

123 People who hadn't taken their dogs to obedience classes were about three and a half times more likely to surrender the animal to a shelter than owners who had: "Owner education could save unwanted pets," Purdue News Service (August 9, 1996).

Page 138 "What strained and anxious lives dogs must lead, so emotionally involved in the world of men": Ackerley, J. R., *My Dog Tulip* (New York: Poseidon Press, 1965), p. 158.

144 "Are you gonna eat that? Are you gonna eat that?": Shepard, Karen, "Birch," From Hempel, Amy & Shepard, Jim (eds.), *Unleashed: Poems by Writers' Dogs* (New York: Crown Publishers, 1995), p. 30.

147 Brian Kilcommons and Sarah Wilson write about a family who projected themselves right out of their own home during feeding time: Kilcommons, op. cit., p. 232.

149 The reaction among men to neutering a dog: Beck & Katcher, op. cit., p. 238.

150 Forty-eight percent of respondents to a University of Pennsylvania questionnaire reported that they saw their pets as "people" rather than dogs: Cusack, Odean, *Pets and Mental Health* (New York: The Haworth

Press, 1988), p. 14.

165 "I will notice that the other two stop and wait for their companion to return. . . . This act is surely indicative of compassion": Masson, Jeffrey Moussaieff, *Dogs Never Lie About Love* (New York: Crown Publishers, 1997), p. 97.

166 Magazine article about a seminar on the practice, which attracted a standing-room-only crown at a pet-care expo in New York: "At the Natural Petcare Expo, talk is cheap," *New York Magazine* (March, 17, 1997).

167 Animals "eat without pleasure, cry without pain, act without knowing it" — Nicholas de Malebranche: From Coren, Stanley, *The Intelligence of Dogs: A Guide to the Thoughts, Emotions, and Inner Lives of Our Canine Companions* (New York: Bantam Books, 1994), p. 64.

OUR DRAMAS, OUR DOGS
Page 182 Story of a young woman who went to her vet with lacerations all over her abdomen and thighs: Beck &

FAMILY DOG

Page 231 Degrees and frequency of "stroking" of dogs among family members: Cain, op. cit., p. 8.

239 Close to a third of participants in a study of 122 families felt closer to the dog than to anyone else in the family: Barker, Sandra B. & Barker, Randolph T., "The human-canine bond: Closer than family ties?", *Journal of Mental Health Counseling*, Vol. 10, No. 1 (January, 1988), pp. 46–56.

243 Married couple who connected profoundly, but solely over their dog: Entin, A. D., "The pet-focused family: A systems theory perspective," paper presented at the Annual Convention of the American Psychological Association, Toronto, Ontario, Canada (August, 1983).

SURROGATE DOG

Page 253 People in the company of a dog are more likely to be regarded by others as friendlier, happier, more relaxed, and less threatening than people who are dogless: Lockwood, Randall, "The influence of animals on social perception," in Beck, Alan & Katcher, Aaron (eds.), *New Perspectives on Our Lives With Companion Animals* (Philadelphia: University of Pennsylvania Press, 1983), pp. 64–71.

254 Dog walkers in public parks and gardens had higher numbers of positive interactions and more extensive conversations with others than people who were either on their own or with small children: Messent, P. R., "Social facilitation of contact with other people by pet dogs," ibid., pp. 37–46.

255 There is a belief "that pets are no more than substitutes for so-called 'normal' human relationships": Serpell, James, *In the Company of Animals: A Study of Human-Animal Relationships* (Cambridge: Cambridge University Press, 1986), p. 24.

257 People with affectionate attitudes toward their dogs have proportionately affectionate attitudes toward people: Brown, L. T., Shaw, T. G. & Kirkland, K. D., "Affection for people as a func-

356

tion of affection for dogs," *Psychological Reports*, Vol. 31, 1972, pp. 957–8.

257 People who interact frequently with their dogs have a higher desire for affiliation with other people than non–dog owners: Serpell, op. cit., p. 38.

257 Elderly pet owners as more self-sufficient, dependable, helpful, optimistic, and socially confident than non–pet owners: Kidd, A. H. & Feldman, B. M., "Pet ownership and self-perceptions of older people," *Psychological Reports*, Vol. 48, 1981, pp. 867–75.

269 Twenty-five million child- rearing American couples: *Current Population Survey*, US Census Bureau, March, 1997.

THERAPY DOG
Page 295 Boris Levinson, an American child psychiatrist, coined the phrase pet therapy in 1964: Levinson, B. M., "Pets: A special technique in child psychotherapy," *Mental Hygiene*, Vol. 48, 1964, pp. 243–8.

296 Implementation of the first pet-facilitated therapy program at a psychiatric unit: Corson, S. A., Corson, W. L., Gwynne, P. H. & Arnold, E. L., "Pet dogs as nonverbal communication link in hospital psychiatry," *Comprehensive Psychiatry*, Vol. 18, 1977.

296 A 1981 study in Melbourne, Australia, evaluated the influence of pets on morale and happiness among nursing home residents: Salmon, I. M. & Salmon, P. W., "A dog in residence: A companion-animal study undertaken at the Caulfield Geriatric Hospital," Melbourne: Joint Advisory Committee on Pets in Society, 1982.

297 Depressed patients in nursing homes have become more interactive and optimistic when visited by dogs and cats: Brickell, C. M., "Depression in the nursing home: A pilot study using pet-facilitated therapy." In Anderson, et al., *The Pet Connection*, pp. 407–415.

297 Prison inmates allowed to take care of animals have become less isolated,

less violent, more responsible, and have exhibited increased morale: Lee, D., "Pet therapy: Helping patients through troubled times," *California Veterinarian*, Vol. 37, 1981, pp. 24–5.

297 Visits by dogs and cats have helped ease feelings of fear, despair, loneliness, and isolation among terminally ill cancer patients: Muschel, I. J., "Pet therapy with terminal cancer patients," *The Latham Letter*, Fall, 1985, pp. 8–15.

297 The presence of a dog at a psychiatric halfway house has helped residents become more social and more adept at communicating: Allen, L. D. & Burdon, R. D., "The clinical significance of pets in a psychiatric community residence," *American Journal of Social Psychiatry*, Vol. 2(4), 1982, pp. 31–41.

297 Populations that have benefited from the presence of animals include elderly veterans . . . : Robb, S. S., Boyd, M. & Pristash, C. L., "A wine bottle, plant, and puppy: Catalysts for social behavior," *Journal of Gerontological*

Nursing, Vol. 6(12), 1980, pp. 722–28.

297 . . . emotionally-disturbed and learning-disabled children . . . : Cusack, op. cit., pp. 96–102.

297 . . . and troubled inner city kids: Beck & Katcher, *Between Pets and People*, p. 48.

301 Petting a dog — in some cases, even being in the same room as a dog — has a calming effect on people: Cusack, op. cit., pp. 64–5.

308 Both men and women in the waiting room of a veterinary clinic petted their dogs for similar lengths of time and with little difference in frequency: Beck & Katcher, op. cit., pp. 88–89.

309 "A Rogerian analyst is not unlike a Labrador retriever": ibid., p. 92.

310 "Dogs," wrote Jerome K. Jerome, "never talk about themselves but listen to you while you talk about yourself, and keep up an appearance of being interested in the conversation": From

Winokur (ed.), *Mondo Canine*, p. 3.

310 "The difficult art in therapy is achieving a mutual feeling of intimacy without touching": Beck & Katcher, op. cit., p. 92.

311 Childhood memory of being in the living room, the staid and sober adults engaged in an elevated discussion about morality and religion while the family's Great Dane lay by the fire . . . : ibid., p. 192.

312 Dogs are like four-legged embodiments of the human id: ibid., p. 183.

312 "We are drawn to dogs," wrote George Bird Evans: From Winokur (ed.), *Mondo Canine*, op. cit., p. 10.

329 Constance Perin, a cultural anthropologist, suggests that the depth of our affection for dogs . . . : From Fogel, Bruce (ed.), *Interrelations Between People and Pets* (Springfield, Illinois: Charles C. Thomas, 1981), pp. 68–88.

330 "In no human relationship can we recreate this symbolic vessel of excess

love . . .": Woloy, Eleanora M., *The Symbol of the Dog in the Human Psyche: A Study of the Human-Dog Bond* (Wilmette, Illinois: Chiron Publications, 1990), p. 20.

EPILOGUE

Page 344 "I feel the same spiritual comfort holding a leash . . .": Conant, Susan, *A New Leash on Death* (New York: The Berkley Publishing Group, 1990), p. 16.

ACKNOWLEDGMENTS

With deepest gratitude to Susan Kamil of The Dial Press and to Colleen Mohyde of The Doe Coover Agency for their unwavering confidence and enthusiasm.

With love to Gail Caldwell, whose friendship, wisdom, and depth truly grace these pages; to Mark Morelli, for so generously attending to both Lucille and me; and to Sandra Shea, Tom Duffy and Rebecca Knapp for providing so much constancy and support.

With sincere thanks to Hope Michelsen, Wendy Sanford, Polly Attwood, Kathy de Natale, Catherine Fabio, and Katie Clark, all of whom have helped make the world of dogs such a rich place. Thanks also to Leslie Hermsdorf of The Dial Press, and to Doreen Manning, for providing invaluable technical and research assistance; to Michael Ian Kaye, Melissa Hayden, and Brian Mulligan for bringing to this project such a sensitive eye for design; and to James Serpell of the University of Pennsylvania for his thoughtful insights about the history of the man-dog relationship.

And with appreciation and affection, as always, to David Herzog.